The Lives of Great Engineers of Ulster
Volume 1

By Professor Sir Bernard Crossland
And
John S. Moore

© 2003

Published by
NE Consultancy for
Belfast Industrial Heritage Ltd.
c/o H & W Old IT Building
The Old Titanic Quarter
Queen's Road, Queen's Island
Belfast BT3 9DT, N. Ireland

Telephone: Belfast: +44 (028 90) 455325
Fax. Belfast: +44 (028 90) 469993
e.mail: kathleen&belfastindustrialheritage.org
Web Site: www.belfastindustrialheritage.org

ISBN 0 9526445 8 4

First published 2003 by NE Consultancy on behalf of Belfast Industrial Heritage Ltd.

ISBN 0 9526445 8 4

Copyright © Bernard Crossland & John S. Moore

The right of Bernard Crossland & John S. Moore to be identified as the authors of these works has been asserted by them in accordance with the Copyright, Designs and Patents Act 1988.

All rights reserved. No part of this publication may be reproduced, stored in or introduced into a retrieval system, or transmitted in any form, or by any means (electronic, mechanical, photocopying, recording or otherwise) without the prior written permission of the publisher. Any person who does any unauthorized act in relation to this publication may be liable to criminal prosecution and civil claims for damages.

Printed by Priory Press, Holywood, Co. Down, BT19 9JE, N. Ireland.

Belfast Industrial Heritage Limited's web site address is www.belfastindustrialheritage.org.

"I warmly welcome your desire to know something of the doings and sayings of the great men of the past and your nation in particular."

The Venerable Bede 673-735

SPONSORSHIP

Belfast Harbour Commissioners
Invest Northern Ireland
Howden Power Limited
Uprite Structures and Services Ltd.
Dr. Gordon S. Millington, OBE.
McKenzies (NI) Limited
Professor John McCanny, C.B.E., F.R.S., F.R.Eng., School of Electrical and Electronic Engineering, The Queen's University of Belfast.

To our wives **Audrey Elliott** and **Oonagh**

BERNARD CROSSLAND

Professor Sir Bernard Crossland, C.B.E., F.R.Eng., F.R.S., left school in 1939 and took up a trade apprenticeship in Rolls Royce Ltd. By part-time and a period of full- time study he graduated in 1943.

After a period as an Experimental Officer in Rolls Royce he left in 1946 to take up an Assistant Lectureship in Mechanical Engineering in the University of Bristol. He had progressed to Senior Lecturer by 1959 when he left to take up the post of Professor and Head of Department in the Queen's University of Belfast. After serving periods as Dean of Engineering and Senior Pro-Vice-Chancellor he retired early in 1984.

In retirement he has been involved in the investigations of several major disasters including the King's Cross Underground fire in 1987. He has served as the National President of the Institution of Mechanical Engineers 1986-87 and of the Welding Institute 1995-98. He has also served on numerous Northern Ireland and United Kingdom governmental committees.

JOHN S. MOORE

After graduation from Queen's University, Belfast, in 1965 as a Mechanical Engineer, he undertook a Graduate Apprenticeship in Rolls Royce Aero Engine Division in Scotland. Following further industrial experience there, he returned to Northern Ireland in 1972. In the Industrial Science Division of the Department of Commerce, he gained a wide knowledge of manufacturing industry throughout Northern Ireland.

There he undertook a study of the Motor Industry in the province which led to a Masters Degree and his book "Motor Makers in Ireland". He was appointed Keeper of the Department of Transport in the Ulster Folk and Transport Museum in 1981. He supervised the world famous collection and the presentation of the history of transport. Following the opening of the new galleries at Cultra he was appointed Administrator. In December 2000 he took early retirement. Since then he has been writing and undertaking consultancy work on all aspects of heritage.

CONTENTS

Author's Note	vii
Acknowledgements	viii
Preface	x
Andrews, Thomas 1873-1912	1
Beatty, James, 1820-1865	8
Brown, John, 1850-1911	16
Cameron, Robert Rupert Gibson, 1903-1979	22
Chambers Brothers	28
(Robert Martin Chambers 1865-1949	
John Henry Chambers 1867-1937 and	
Charles Edward Chambers 1873-1932)	
Dargan, William, 1789-1867	36
Davidson, Sir Samuel. 1846-1921	41
Dunlop, John Boyd, 1840-1921	48
Ferguson, Harry, 1884-1960	55
Harland, Sir Edward, 1831-1895	65
Lanyon, Sir Charles, 1812-1889	71
Mackie, James Jnr., 1864-1943	78
Martin, Sir James 1893-1981	84
Megaw, Eric Christopher Stanley, 1908-1956	92
Mitchell, Alexander, 1780-1868	100
McCandless, Rex, 1915-1992	107
Pirrie, William James - Viscount Pirrie of Belfast, 1847-1924	115
Pounder, Cuthbert Coulson, 1891-1982	124
Rebbeck, Sir Frederick, 1877-1955	132
Stevenson, John, 1850-1931	143
Thomson, James, 1822-1892	149
Thomson, William; Lord Kelvin of Largs, 1824-1907	157
Traill, William Acheson, 1844-1933	166
Wolff, Gustav Wilhelm, 1834-1913	173
Workman, Francis, 1856-1927	178
Appendix 1 Index to and source of illustrations.	183

AUTHOR'S NOTE

In quoting ship tonnages there are two alternatives, which are commonly used. In this book the tonnages quoted are gross tons, based on the volume of the space in the vessel, which is more frequently used. The alternative, which is used more frequently by the Ministry of Defence, is displacement tonnage; this is the weight of water displaced by the ship.

ACKNOWLEDGEMENTS

Professor J. A. Carson Stewart lately **Head of the School of Electrical and Electronic Engineering** of the **Queen's University of Belfast** for contributing the essay on Eric Christopher Megaw (p.92) which is much appreciated.

Many people have been most helpful to us in writing this book. In particular we would like to record the help we have received from **Mrs Kathleen Neill, Executive Director of Belfast Industrial Heritage Ltd.** in editing and producing the book. **Ms. Jennifer Crossland, Reference Librarian** in the **Belfast City Library** has been a font of knowledge of sources of information and illustrations. **Mr. Ken Anderson, Senior Photographer** of the **Ulster Folk and Transport Museum** has located and provided copies of some of the photographs reproduced in this book.

In writing this book our task has been greatly aided by contact with some of the descendants of those we have written about. In particular we would like to acknowledge **Mr. Robin Cameron**, son of Rupert Cameron, **Mrs. N. Bleakley**, granddaughter of Sir Samuel Davidson; **Mr. Gordon Mackie**, the grandson of James Mackie; **Mrs. V. Pounder**, daughter-in-law of Cuthbert Pounder; **Mrs Maureen Foxall**, daughter of Sir Frederick Rebbeck; **Mr. Raymond H. Hall**, grandson of Sir Frederick Rebbeck. **Mr. Terence Chambers**, grandson of Charles Chambers; and the late **Mrs. Elizabeth (Betty) Sheldon**, daughter of Harry Ferguson.

Mr. W.R.A. Stafford, recently retired **Overseas Director of Howden Sirocco**, has been most helpful in providing information on Sir Samuel Davidson and Davidson and Co. Ltd. now Howden Sirocco, and of the recent development of their Chinese company. **Ms. S. Harbinson** of **Langford Lodge Engineering Company Ltd.** and **Mr. Dei Hoyland** of **Martin-Baker Aircraft Company Ltd.** have provided photographs and information relating to Sir James Martin. **Mr. Bryan McCabe** of **W. & G. Baird**, which took over McCaw, Stevenson and Orr, has provided information on that company. The late **Mr. J. H. McGuigan** provided much information on the Giant's Causeway Tramway. **Professor Adrian Long**, at present **President of the Institute of Civil Engineers**, gave us reference to Sir Samuel Davidson's design of the ventilation system of the Royal Victoria Hospital, and also suggested the name of James Beatty as the subject of one of our essays. When talking with **Gordon Millington**, lately **President of the Irish Academy of Engineering**, he suggested the name of Alexander Mitchell, the Blind Engineer of Ulster.

Lastly, one of the authors, BC, would like to record his gratitude to the School of Chemical Engineering in the Queen's University of Belfast for providing office accommodation and a welcome into their midst. And especially the invaluable assistance so willingly given by **Ms. Muriel Orr** and that of **Mrs Nora Ferguson**, School of Civil Engineering, **Queen's University of Belfast**.

PREFACE

There appears to be a lack of appreciation of the great contribution of Ulster, particularly Belfast, to engineering manufacture. Belfast was one of the leading centres of manufacture in the British Isles in the nineteenth and first half of the twentieth centuries. Few people appear to be fully aware of the extent and scale of the engineering innovation exhibited by this small province. Some innovation came from the innate abilities of Ulstermen while some came from people attracted from elsewhere by the high level of innovation and dynamism in the province.

The purpose of this book is to draw together twenty-five examples of great innovative engineers of Ulster, though no doubt other names will come to the minds of many readers. We hope that the next generation will be enthused to carry on or revive the great tradition of innovation and entrepreneurship. It may also provide educationalists with thoughts on the best way to draw out these attributes in their students.

We had to come up with a definition of the category of individuals to include. Firstly we decided, with one exception, that they should have been dead for at least twenty-five years to enable us to gain a clear perspective of their contribution. In essence those selected were either born in Ulster or made their major contribution to their life's work in the province. For instance, William Dargan made a significant contribution to the commercial wellbeing of Belfast through the dredging of the channel of the River Lagan and the creation of Dargan's Island (now Queen's Island) as well as constructing the railways which served the city. While Eric Megaw is probably less well known for his development of the compact high power cavity magnetron, used in airborne radar in the Second World War, which subsequently has been applied to the microwave oven.

Many of the great engineers we have included had, by the standards of their day, a good education. The privileged few were privately educated at home, while many in the Belfast area were fortunate to be educated at the Royal Belfast Academical Institution, RBAI. The RBAI was founded in 1814 with both school and college departments until sometime after 1845 when the college department was absorbed into the newly created Queen's College of Belfast. The school developed an excellent reputation and it attracted some high calibre teachers such as James Thomson, the father of Lord Kelvin, and the Rev. Osborne Reynolds the father of the world renowned engineer Osborne Reynolds. The school provided an excellent broadly based foundation mainly for children of the middle class who entered the professions or were the mainstay of commerce and the blossoming engineering companies. Clearly a good general education formed an excellent foundation on which to build.

The examples of great engineer we have considered give some food for thought on the education and training of engineers which may have a relevance to the present day. Some of these engineers, for instance Thomas Andrews and Rupert Cameron, have drawn attention to the crucial importance of concurrent education and training. This not only ensured that they recognised the relevance of theory to practice but as importantly they got to know and respect people on the workshop floor. They recognised that such experience was vital for those who aspired to and reached managerial positions where man management was important. The lack of such experience has too often been responsible for the industrial strife which has led to the decline of our manufacturing base.

Those who obtained further education after leaving school, such as Sir Frederick Rebbeck, Rupert Cameron and the Chambers brothers, achieved it by evening school study after a long working day; what some refer to as the hard way. Modern society in general finds this unacceptable and with the exponential growth of knowledge full time education has become a prerequisite. But this does not negate the need for a well-considered period of training in industry before, during or after full time education. Unfortunately, with the decline in our manufacturing industry and particularly our larger companies, there are not the training places needed nor in many companies is there the will. The vacillation of our Government has hardly helped.

It should not be forgotten that the success of these great engineers has depended on the loyal support of their staff and workforce and the support of their wives and family.

THOMAS ANDREWS Jnr.

Thomas Andrews Jnr. was born in Comber, County Down, the second son of a family of four boys and one girl, on the 17 February 1873. His father Thomas Andrews of Ardara came from a prominent business and political Ulster family. His ancestors had settled in Mahee Island, Co Down in the early seventeenth century, and in the late seventeenth century they became established in Comber. There they created various businesses culminating with the Andrews flax spinning mill built in 1864 which continued in production up till 1997. His mother, Eliza Morrison Pirrie, was the sister of William James Pirrie, later the 1st Viscount Pirrie, Chairman of Harland and Wolff 1896-1924.

As a young boy Thomas developed a love of boats, horses, country pursuits and sports. He was educated by a tutor at home until 1884 when he became a student at the Royal Belfast Academical Institution. At school he excelled at cricket and other sports but showed no special aptitude for study. He was highly regarded as a well principled and friendly human being by both his fellow student and teachers.

On the 1 May 1889 he left school to start a premium apprenticeship in Harland and Wolff (H&W). Though he was well connected with senior management he received no privileges nor did he expect them. H&W were at the beginning of a unprecedented development of ever larger ocean liners, in particular for the White Star Line. The premium apprenticeship gave Thomas experience in many departments finishing up with eighteen months in the drawing office. Work began at 6 o'clock and in the evening Thomas pursued studies in such subjects as machine drawing and sketching, applied mechanics and naval architecture.

This left him with a strong belief in the importance of technical education and also in the necessity of practical knowledge gained in workshops. He also recognised the importance of working with, and developing an appreciation of the skills and the contribution of, the workforce when it came to managing them. During his apprenticeship he earned the respect and lasting friendship of all of those he worked with which was to stand him in good stead in his meteoric career.

Towards the end of his apprenticeship Thomas was entrusted with responsible duties. Whilst he was still twenty he supervised the construction work on the 726 ton cross-channel *Mystic* and he successfully represented the company on the trials of the 7,669 ton White Star liner *Gothic*. Having completed his apprenticeship from May 1894 through to 1896, he assisted the joint Managing Directors appointed by W.J. Pirrie, R.M. Carlisle the General Manger and J. Bailey the Commercial Manager.

From 1896 to 1900 he was first Outside Manager and then Head of the Repair Department which greatly widened his experience. One of his more demanding tasks, which attracted considerable interest in the technical press, was the lengthening of the *Scot*. This involved cutting it into two and inserting a new section. He was also responsible for the extensive repair of the 7,899 ton *China* completed in the yard in 1894. This vessel had been severely damaged when she ran aground on an island in the Red Sea. This experience stood him in good stead when in 1900 he became Manager of Construction at a frenetic time in the construction of ever larger ships for such lines as the White Star, Hamburg Amerika and Holland America. These included the 20,804 ton *Celtic*, the 21,073 ton *Cedric* and the 23,175 ton *Baltic*.

In 1904 Thomas became Assistant Chief Designer and shortly after, in 1905, Chief Designer. In this position he had responsibility for the design of many well known ships, including the famous *Olympic* and *Titanic*, the greatest ships of their day. In July 1907, over after dinner drinks and cigars, Lord Pirrie, Chairman of the H&W and his friend J. Bruce Ismay, Chairman of the White Star Line, discussed the competition posed by Cunard. Their magnificent turbine powered flagship, the 31,900 ton *Mauritania*, had recently gained the Blue Riband for the fastest crossing of the Atlantic. At the conclusion of that evening they had envisaged the building of three liners which would be the largest and most luxurious ever to be built and, if needs be, capable of capturing the Blue Riband. They even came up with names *Olympic*, *Titanic* and *Gigantic* - though the latter when built became the *Britannic*. The purpose was to ensure a regular transatlantic service in each direction in luxurious surroundings.

Thomas Andrews with his Assistant Chief Designer, Edward Wilding, were responsible for the design of these 46,000 ton liners while Alexander M. Carlisle, Managing Director of the Shipyard, was responsible for their construction. To accommodate the building of these liners Sir William Arrol and Company, who had been responsible for the building of the Forth Railway Bridge, completed in 1890, built the massive gantry above nos. 2 and 3 slipways, which dominated Queen's Island for many decades. In 1905 Lord Pirrie had convinced the Belfast Harbour Commissioners to undertake the construction of the Thompson graving dock, large enough to accommodate these liners. So, all was in place to construct their massive ocean liners.

Titanic's keel being laid on No. 3 Slip in the Arrol Gantries

The order for the *Olympic* and *Titanic* was placed in 1908. The engine arrangement was the first consideration and Thomas adopted that used for the first time in the 14,802 ton SS *Laurentic* completed in 1909. The arrangement consisted of a pair of massive four cylinder triple expansion engines driving the port and starboard propeller shafts each delivering 15,000 hp at 75 rpm. These two engines exhausted into a Parsons direct coupled steam turbine driving the central propeller shaft and delivering 16,000 hp at 165 rpm. As the turbine could not operate in reverse, it had to be bypassed when the vessel was being manoeuvred, by diverting the exhaust steam from the reciprocating engines direct to the condenser.

Left: Triple Expansion Engine for the Port and Starboard Propeller shafts. Note the man standing on the bed plate. Right: Steam Turbine to drive the central propeller shaft.

The *Olympic* and *Titanic* had double bottoms with longitudinal girders. Fifteen transverse watertight bulkheads extended from the double bottom to the upper deck, 11ft above the water line, at the forward end, and to the next deck, the saloon deck, at the aft end. That provided for any two compartments to be flooded without compromising the safety of the ship. This more than satisfied the Board of Trade regulations. However, additional safety could have been achieved by extending the bulkheads up to the first continuous watertight deck - the shelter deck one up from the saloon deck.

The keel of the *Olympic* was laid down in December 1908, followed by the *Titanic* in March 1909. The *Olympic* was launched on 20 October 1910 and the *Titanic* on 31 May 1911. After sea trials the *Olympic* sailed on its maiden voyage from Southampton to New York on 14 June 1911. After a collision with the HMS *Hawke* in the Solent on 20 September, she returned to Belfast where she displaced the *Titanic* from the Thompson dock to allow repairs to her plating. In March 1912 she again returned to Belfast having lost a propeller blade in the North Atlantic, but she returned to service before the departure of the *Titanic*. She was modified after the *Titanic* disaster and became known as 'The Old Reliable.' As a result of the economic depression *Olympic* finally went to the breakers yard in 1935 not long after a complete overhaul.

The *Olympic* (left) and *Titanic* (right), March 1912.

In the meanwhile, *Titanic* having been fitted out and having successfully completed her sea trials, departed for Southampton on 2 April 1912. Thomas Andrews was on board and planned to sail on her maiden voyage to New York. Andrews probably replaced Lord Pirrie, who was seriously ill, otherwise Lord Pirrie might have accompanied J. Bruce Ismay. The *Titanic* left Southampton on her maiden voyage to New York on 10 April.

At about 10.40 pm on 14 April the lookout saw an iceberg ahead and warned the bridge. The First Officer ordered the wheel hard-a-starboard, instructed the engine-room to go full speed astern and pressed the button which closed the watertight doors in the bulkheads. The ship, which had been travelling at 22.5 knots, had lost little speed when it struck the iceberg a glancing blow.

Most people only felt a jar and did not recognise its significance. Thomas Andrews was at work in his cabin when he was called by the Captain. They carried out a quick inspection and discovered that the first five compartments were badly holed and the water level rapidly rising. Andrews realised the ship was doomed. As these compartments filled the ship would sink lower in the water until the water in the fifth compartment overtopped the bulkhead. This would create a domino effect with successive compartments becoming flooded. He estimated 2 hours before it sank; it remained afloat for 2 hours 25 minutes.

Survivors recorded that Thomas Andrews remained completely calm and collected, doing everything he could to help passengers and crew. He calmly persuaded passengers to get into the lifeboats when many of them thought the *Titanic* was unsinkable and it was best to stay on board. Later he was seen throwing deckchairs overboard to those struggling in the water. There is still debate on the precise numbers of survivors and those lost but the accepted figures are 705 and 1,509 respectively.

The whole world wanted to know how the *Titanic*, which was supposedly unsinkable, had come to sink? Why did the bulkheads not extend up to the shelter deck, why were there insufficient lifeboats to accommodate all those on board and why, in the event, were so many lifeboats only partially filled? It was easy to blame the designer who had gone down with the ship. The truth was that designers at that time, and many decades after, worked to rules laid down by the Board of Trade. The height of the bulkheads and the number of lifeboats more than satisfied the rules. The Board of Trade Inquiry into the loss of the *Titanic*, held in May 1912, completely exonerated H&W and praised the quality of workmanship but it was critical of the Board of Trade Rules.

Like so many disasters before and after the loss of the *Titanic*, the real fault was a failure of management in this case Captain Edward Smith in command. Earlier in the day he had received warnings of the presence of icebergs by radio from ships in the vicinity. Despite this he continued on at full speed even in hours of darkness; surely gross irresponsibility? Perhaps he was influenced in his decision by the presence of Bruce Ismay, the Managing Director of the White Star Line, on board. The fact that only 705 were rescued when the lifeboats had capacity for 1,178 surely indicates that the crew were inadequately trained in lifeboat drill and evacuation procedures? None of this reflects on Thomas Andrews who appears to have behaved in an exemplary way up to the moment of his death.

At the beginning of the twentieth century Thomas Andrews was elected a Member of the Institution of Naval Architects, the Institution of Mechanical Engineers and the Society of Naval Architects and Marine Engineers (New York) and an Honorary Member of the Belfast Association of Engineers. In 1908 he married Helen Reilly Barbour and they settled in Dunallan, Windsor Avenue, Belfast. In 1910 they had a daughter. His widow remarried in 1917 to Henry Pierson Harland a nephew of Sir Edward Harland.

Sources:
1. 'Thomas Andrews - Shipbuilder' by Shan F. Bullock, reprinted the Blackstaff Press, 1999.

2. 'Shipbuilders to the World - 125 years of Harland and Wolff, Belfast 1861-1986' by Michael Moss and John R. Hume, pub. The Blackstaff Press, 1986, ISBN 0 85640 343 1.

3. 'Anatomy of the Titanic' by Tom McCluskie, pub. PRC Publishing Ltd., 1998, ISBN 0 856482 3.

4. 'A Night to Remember' by Walter Lord, pub. Penguin Books, 1978.

5. 'A Taste of Old Comber - The Town and its History' by Len Ball and Desmond Rainey, pub. The White Row Press, 2002, ISBN 1 870132 06 8.

JAMES BEATTY

James Beatty was born on the 31 March 1820 in Enniskillen the son of Doctor James Beatty. He was educated by his father in classics and by a tutor in English and Mathematics. He showed such mathematical and constructive skills, as well as being interested in the newly developing railways, that his father determined he should become a Civil Engineer. At the age of fifteen he became a pupil of Mr T.J. Woodhouse, M.I.C.E. who was involved in the construction of railways. Woodhouse had previously been the County Surveyor for Co. Antrim and he had been succeeded by Charles Lanyon, later Sir Charles.

Though Richard Trevithick had built the first recognisable steam locomotive in 1803, it was not until the 27 September 1825 that the first recognisable railway, the Stockton to Darlington Railway, was opened. This ran for twenty-six miles from coal mines near Bishop Aukland to Darlington and then on to Stockton on the navigable estuary of the Tees. At its western end it was cable hauled by stationary steam engines over two ridges with a section in between worked by horses. Then for most of the rest of the distance it was steam locomotive hauled. The railway was used to transport coal but passengers and parcels were carried by horse drawn carriages.

The first railway to be built which envisaged a combined cable and steam locomotive hauled regular passengers as well as goods service was the less well known Canterbury and Whitstable Railway opened on the 3 May 1830. This was followed by the much more widely known Liverpool and Manchester Railway opened on the 15 September 1830 which employed mainly steam locomotive hauled trains. The success of these enterprises created the explosion of interest in building railways throughout the United Kingdom and worldwide. This was the world which James Beatty entered when he started his pupilage in 1835.

In 1835 Woodhouse was the Resident Engineer for the construction of the Midland Counties Railway. Beatty became so indispensable that at the end of his pupilage, at the age of nineteen, he was appointed as Assistant Engineer on the Railway. In 1840 he was engaged to survey the Manchester and Derby line to establish the most economic route through the Derbyshire High Peak.

Beatty joined Messrs Peto and Betts on the Norwich and Lowestoft Railway in 1842. Peto and Betts, later Peto, Brassey and Betts, was one of the great railway contractors of the Victorian age. In 1845, still only twenty five, Beatty was appointed as their Chief Agent and Engineer in charge of the execution of works on the Southampton and Dorchester Railway.

Peto, Brasey and Betts were involved in railway construction worldwide and in 1852 they dispatched Beatty, with Donald Campbell in support, to explore and lay out the European and North American Railway in New Brunswick. This railway was planned to connect Halifax, which was the closest port to the United Kingdom, to the interior of Canada and the United States. Within a little over three months, as a result of Beatty's great energy and leadership, they laid out the routes, many of them through primeval forest, and produced estimates of the necessary works before returning to England in late autumn.

In the following spring he was dispatched to Nova Scotia to explore and lay out the further section of the European and North American Railway in that province. Many of these routes were again through dense primeval forest through which the survey parties had to cut their way. During this period the survey parties were living in primitive conditions in tents with the most basic rations of salt pork and hard biscuits. As winter approached the temperature sank to minus 5F with three feet of snow. Beatty led from the front and was responsible for co-ordinating the teams which involved travelling between the survey parties and sharing their hardship.

During 1853 trouble erupted between Russia and Turkey over the question of the protection of Christians and the Holy Places in the Levant. France was also in disagreement with the Tzar over the protection of the Holy Places while the British were paranoid about the threat to India posed by Russian expansionism in Asia and the Middle East. These troubles culminated in November with the Russian fleet sinking a squadron of the Turkish Navy with the danger of further escalation. The increasing fears of a major war unsettled the financial supporters of the European and North American Railway. This resulted in construction work on the Railway being halted and the return of Beatty and his staff to England in February 1854.

On the 22 March 1854, France and Great Britain declared war on Russia who had invaded the Turkish provinces in the Balkans. French and British ships sailed through the Bosphorus and forced the Russian Fleet to take refuge in the Port of Sebastopol in the Crimea. British troops together with those of their allies landed in the Crimea on the 14 September and by the end of the month they were encamped on the high ground south of Sebastopol where they set up siege lines and batteries. Unfortunately, no provision had been made for the adequate shelter of the besieging troops or for their adequate provisioning, medical services and supply of ammunition. These could only be provided by horse and cart along completely inadequate roads from the small deep water harbour of Balaklava.

The appalling conditions of the besieging troops were exposed by the famous war reporter from the Times newspaper, William Howard Russell. These reports created great public and political concern, which in November prompted Samuel Morton Peto MP to propose to the Secretary of War, the Duke of Newcastle, that a railway should be constructed between the deep water harbour at Balaklava and the front. Within a surprisingly short time, on the 2 December, Peto, Brassey and Betts were given the go ahead. It was a formidable task as time was of the essence and it attracted great public interest.

Peto had been impressed with James Beatty's powers of leadership so he appointed him as Engineer in Chief for the company in the Crimea at what was then the princely salary of £1500, with Donald Campbell as an assistant engineer. A surveying party under Campbell was dispatched to the Crimea early in December to survey the route of the railway. Beatty's immediate task was, with Edward Betts, to organise the extensive supplies not only to build and equip the railway but as importantly to house and support the substantial workforce and stabling for the horses needed. He became involved in recruiting the first 250 navvies as well as a surgeon and sought the advice of Thomas Brassey on constructional details.

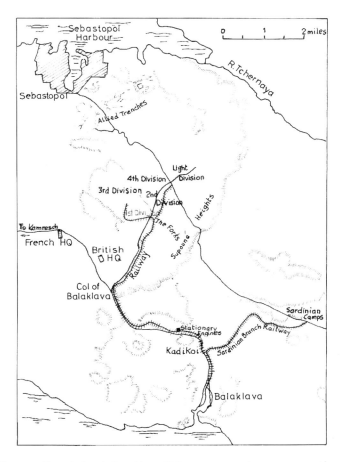

The force dispatched to the Crimea was known as the Civil Engineering Corp but it was intended that they would work only in a civilian capacity. In a surprisingly short time the first group of navvies with some construction equipment was dispatched in a clipper on the 21 December. This was shortly followed by the remaining eight ships, seven steamers and another clipper which formed the first flotilla. These contained thousands of tons of equipment and supplies, horses, and the main contingent of navvies. James Beattie spent Christmas at home before departing with three assistants via Marseilles to arrive at Balaklava on the 19 January 1855 prior to the arrival of the first ship.

Beatty was welcomed by Campbell and his survey party who had carried out a preliminary survey of the route of the railway. The proposed route from Balaklava followed the gorge to the village of Kadikoi, a distance of 1½ miles, where there was sufficient level space to construct sidings and sheds. From Kadikoi the route climbed over 500 ft to the plateau where the Army Commander, Lord Raglan, had his headquarters. The initial incline from Kadikoi included an incline of 1 in 14, too steep for a steam locomotive or for horse drawn wagons. It was proposed to install a stationary winding engine at the top of that incline and two stationary winding engines had been included in the equipment transported to the Crimea. The rest of the incline was suitable for horse drawn wagons. It was proposed to build a terminal adjoining Raglan's headquarters with branch lines to the individual camps of the besieging army. All this had to be constructed within a few weeks during the bitter Crimean winter.

On his arrival Beatty and his three assistants sought assistance from the army prior to the arrival of the first vessel with navvies. This assistance allowed them to get started to clear the ground adjoining the wharf and to clear a way through the town for the railway. Early in February the flotilla carrying the navvies began to arrive and besides extending the rail-track towards Kadikoi they also erected their accommodation huts on the hillside overlooking the town and harbour. By the 21 February they had completed the track up to Kadikoi and laid the rails. They had also extended the sidings and erected sheds adjoining the wharf, as well as unloading one of the stationary winding engines and erecting and testing it.

The commencement of the railway at Balaklava.

The Pre-fabricated Huts for the Railway Navvies.

The completion of a single line to Kadikoi enabled horse drawn wagons to be used to create an ammunition and supply depot adjoining the village and the construction of sheds and sidings. In the meanwhile, the rail-track was being driven from Kadikoi up the incline to the flagstaff on the plateau adjoining Raglan's headquarters. The stationary engine at the top of the 1 in 14 incline was installed. By the 24 March the railway to the summit, a distance of $4\frac{1}{2}$ miles, was in operation. In addition a second track had been laid between Balaklava and Kadikoi and a branch line laid to the Diamond Wharf on the opposite side of the deep water inlet to Balaklava.

On the 2 April sick and wounded soldiers were carried down from the summit in relative comfort in wagons to Balaklava which probably saved many lives. Unfortunately, on the 5 April Lord Raglan requested the return of a regiment to Balaklava. They were divided in three trains which went down the 1 in 24 incline to the stationary engine under gravity using their brakes to control their speed. The last of these trains with Beatty and the officers on board got out of control. In endeavouring to stop it, Beatty was thrown off and received internal injuries. However the seriousness of his injuries was not recognised and he continued in charge till invalided back to England in November 1855.

In April the branch line to the Worontzoff Road to serve the 2nd and Light Divisions, and another branch to the 3rd and 4th Divisions were completed. Later in the year, after the fall of Sebastopol, the

Progress of the Railway at Kadikoi.

branch railway to the Sardinian camp from Kadikoi was constructed. This made the total length of track 14 miles plus a few miles of sidings. Steam locomotives were used on the line from Balaklava to Kadikoi and to the Sardinian camp but they do not appear to have been used from the stationary winding engine up onto the plateau.

The siege of Sebastopol, which ended in its capitulation on the 9 September 1855, was the first example of modern warfare; it was the Stalingrad of the 19th century. Success depended on the deployment of massive fire power and the associated logistics of supply. British Artillery fired 28,476 rounds in the three days of the sixth bombardment at the end of August 1855. These were of various calibre shells. The bulk of these were carried to the front by Beatty's railway. In addition the railway carried the rest of the supplies for the troops, as well as removing the sick and wounded. The success of the Crimean War depended on the logistics of supply and demonstrated the importance of railways in modern warfare.

The success of Peto, Brassey and Betts in organising at very short notice the logistic support for a major civil engineering project, and James Beatty in managing the project to a very tight time-scale, opened the eyes of the Army which was normally resistant to change. It led to the creation of the Land Transport Corp which took over the running of the Crimean Railway from James Beatty. The Land Transport Corp became an integral and important part of the Royal Engineers which took an active part in many subsequent military campaigns worldwide including both World Wars. This was a lasting memorial to James Beatty.

James Beatty stayed on long enough to oversee the refurbishment of the railway substantially completed. But ultimately ill health compelled him to hand over to Donald Campbell and he left the Crimea in November 1856 and returned to London via Scutari on the Bosphorus arriving in December. His condition continued to deteriorate and he died at home on the 11 March 1856 shortly before his thirty-sixth birthday. A post-mortem examination confirmed the diagnosis of an aneurysm of the aorta caused by the accident.

James Beatty's grave in Kensal Green Cemetery.

Beatty's death was greatly mourned by Sir Morton Peto, Thomas Brassey and Edward Ladd Betts who had an imposing memorial erected in Kensal Green Cemetery. Due to the actions of Sir Morton Peto the Government awarded a pension to Beatty's widow as if he had been a Colonel. Both William Russell (the war correspondent for the Times) and Roger Fenton (the famous war photographer) endorsed a view of James Beatty as an admirable and dedicated man. They could talk with authority as they had been present in the Crimea.

Sources:

1. 'Report to Peto, Brassey and Betts' by James Beatty, pub. Illustrated London News, 12th May 1855.

2. 'The Grand Crimean Central Railway' by Brian Cooke, pub. Cavalier Press, 1990.

3. 'Ireland and the Crimean War' by David Murphy, pub. Four Courts Press, 2002.

4. 'James Beatty' pub. Institution of Civil Engineers Memoirs, p 154-158.

JOHN BROWN

John Brown FRS, was the eldest son of John Shaw Brown, the damask linen manufacturer who had premises in Edenderry and Belfast. He was born on 24 February 1850 at Donard Lodge, Waringstown and was educated at the Academical Institution, Belfast and at Bonn University. He served an apprenticeship in the firm and, after going through every department, became a partner.

Soon after the death of his wife in 1882, Brown eased himself out of the management of the family linen business and devoted the rest of his life to scientific and engineering pursuits which he had been pursuing as a hobby to that point. Some were for practical purposes and some were purely investigative with no immediate commercial potential. His first patent 2908 of 1885 was for "Combing and Clipping of Fringes for Doyleys "[sic]. This was a practical application developing the rollers and combs of a carding machine for the specific purpose of straitening and cutting the ornamental fringe on household fabrics such as linen doilies, tablecloths or towels.

The primary battery was the main source of electrical power at this period and he undertook many experiments into the problems of converting chemical activity into electrical energy. As an offshoot of his investigations he was granted patent 6936 of 1885 "Adjustable Photographic Camera Stand", which he seems to have developed in pursuit of his interest in the photographing of electrical discharge phenomena. He seems to have been the first person to publish the aurora like patterns produced on photographic plates by a high voltage electric discharge.

Before the full understanding of the role of the electron in chemical and electrical reactions, Volta, Faraday and Kelvin had produced empirical and somewhat vague reasonings of these phenomena. Brown, having studied the electric potential between two dissimilar metals (zinc and copper) in air and hydrogen sulphide and showed a change in potential between the two surrounding media and so helped move forward the understanding of the basic physical chemistry of batteries. He was a member of the Electrolysis Committee of the British Association. He was of sufficient standing to publish the results of his experiments in the Philosophical Magazine, the Proceedings of the Royal Society and in the Proceedings of the Belfast Natural History and Philosophical Society. His other interests included natural history and education, topics on which he also lectured to the British Association and the Belfast Natural History and Philosophical Society.

Much of his work was done in his own laboratory but he collaborated with J. D. Everett in the Physics Department of Queen's College, Belfast and Professor George FitzGerald of Trinity College Dublin. It was in the latter that he suffered an industrial injury when the chemicals he was mixing exploded and he lost an eye. He had been working on the effect of different media on the voltaic potential and was using heated petroleum. The Health and Safety Rules at the time were so very different from those currently in force.

Jack Brown in a Seprollet Steam Car in 1896

At this period there were great advances in the generation and use of electricity. At the 1879 Berlin Exhibition, the Siemens brothers had demonstrated a working electric railway using their patented dynamo and motor. Other investigators such as Thomas Alvar Edison and Charles Van Depole then staged small demonstrations in America and around the world using electricity to power railways or tramways. There was a considerable interest in creating a vehicle which could roam freely without the need for rails or connection to a permanent power source, in effect a horseless carriage. Brown took an active interest in all forms of transport. In 1896, he was the first man in Ireland to own an automobile, a French Seprollet steam car. In November that year he presented a lecture to the Belfast Natural History and Philosophical Society about "Automobilism" which he illustrated by having the vehicle on stage in the Ulster Hall and demonstrated it in action at the Belfast Waterworks, on the Antrim Road the following day.

In the late eighteen eighties, the two German pioneers, Daimler and Benz, individually developed the internal combustion engined motor car. There developed an argument between the proponents of the different power sources, electricity, steam and the internal combustion engine, as to which was the best for use. Using his experience with batteries, Brown developed his own electric vehicle which he called an "electric street boat" or "electric gondola".

The "Electric Street Boat"

Brown believed that the state of roads in Ireland was a serious obstacle to the introduction of the new form of transport and founded and was President of the Irish Roads Improvement Association. He invented the "Viagraph" to record the deformations of the road surface and patented it under 9123 of 1898. This device was towed by a vehicle and recorded vertical vibrations caused by the imperfections in the road surface. The results were dramatic graphs which highlighted the effect of the various haphazard methods of road repair currently in use. The County Surveyors were not happy with the adverse publicity but did eventually respond to the pressure and adopt methods more suitable to the needs of the motor car. He appears to have had another patent application on the subject, filed as 13952 of 1911, awaiting full details at the time of his death.

As well as attempting to influence the authorities and have them improve the maintenance of the roads Brown adopted an alternative approach. He designed and patented a spring wheel as number 7692 of 1902. In this design he specified the use of tapering coil springs made from thin flat bar as the spokes of the wheel. They were set alternatively in such a way that they resisted the lateral loads involved when cornering. The patent also saw a use for the design in reducing fluctuations in power transmission pulley wheels. In the event, improvements in pneumatic tyres and road surfaces meant there was no need for the complications of this design in road transport.

Patent Spring Wheel

As well as his pursuit of improvements in road transport, Brown took an interest in railways and took out a patent, number 20887 of 1900, "Improvements in the Working of Vehicles or Trains on Railways or the like". In this he envisioned a train consisting of a series of easily connected or disconnected carriages running on a looped circuit. As the train approached a station the last car would be disconnected and be guided into a platform where the passengers would disembark. When the next train was due, the carriage would be accelerated out on to the main line and take the position of lead coach and the main train would connect up. The passengers would then move through the train to the coach which would be detached at their destination. Brown envisaged that the whole system could be operated by remote control. However, if it was deemed necessary a driver could control each carriage in the time from leaving the main train into the station and out again and the driver of the leading coach would have control of the full train between stations.

His passion for electricity continued and was applied to other problems. He developed a battery powered lawn mower for his extensive grounds at "Longhurst", his house on the upper Malone Road in Belfast. He also took out a patent, number 7861 of 1900 "Improvements relating to Electrical Reciprocating Apparatus". He envisaged his invention being used to power all sorts of tools such as a saw, hammer, rock-drill, knife or egg-beater. This employed a simple make and break circuit to a solenoid so that when the circuit was closed, the tool was raised and it fell under the action of gravity. A spiral groove could be employed to provide a partial rotary motion. Also he allowed for a double acting solenoid circuit so that the device could be used in a horizontal position.

Having taken to the internal combustion engine, Brown bought and drove a series of cars as reliability improved, designs changed and engines became bigger and more powerful. From the driving wheel of these cars he experienced the problem of dust thrown up by car tyres from dry roads. He then tried to eliminate the problem by driving a fan from the car engine and sucking the dust laden air from around the wheels into a collector. There the dust was mixed with water to form briquettes which were then dropped along the road. This was covered by patent 19621 of 1903. The idea was first tried out on his "curved dash" Oldsmobile but the relatively small engine was unable to drive the fan and provide enough power to carry the twenty stone Brown up the hill from Shaw's Bridge or on his journeys from home to Queen's University.

Jack Brown was elected a Fellow of the Royal Society in 1902, was an Associate of the Institution of Electrical Engineers and read papers to the British Association. He was local secretary to the British Association when it met in Belfast in 1902. He died 1 November 1911

Sources:

1	Obituary, Proceedings of the Royal Society, vol. 88, London, 1913

2	Belfast Natural History and Philosophical Society, Centenary Edition, pub. Belfast 1933.

3	Belfast Telegraph

4	'Motor Makers in Ireland', John S. Moore, The Blackstaff Press, Belfast, 1982, ISBN 0 85640 264 8.

ROBERT RUPERT GIBSON CAMERON, OBE

Rupert Cameron was born on the 24 October 1903 in Carrickfergus, Co Antrim, the second of three children of Gibson Cameron and his wife Jane, daughter of Henry Beattie of Carrickfergus. His father and an uncle ran a bakery and, from a young age, he was fascinated by the gas engine and its hit-and-miss governor which drove the dough mixers and other machinery and the boiler which supplied steam to the ovens. He also took a lively interest in the early motor delivery vans and, during his early teens, in the school holidays, he delighted in driving the bread servers about on their rounds. When he was a little older he got a motor cycle and his love of motor cycles was to remain with him all his life.

He was educated at Carrickfergus Model (Primary) School and won a scholarship to the well known Royal Belfast Academical Institution. In 1919, shortly before he was 16, Rupert Cameron left school having matriculated to take up a ship draughtsman apprenticeship in Harland and Wolff. His love of ships perhaps derived from his maternal grandfather and great uncle who were both sea captains. Perhaps, having grown up on the shore of Belfast Lough, he was also attracted by the ships which went up and down to the Belfast Harbour as well as the new ships built in the Belfast shipyards. These may well have included the magnificent *Olympic* and its ill-fated sister ships the *Titanic* and the *Britannic*.

At that period access to Higher Education was mainly limited to the children of affluent parents and a very few scholarship holders. The saving grace in 1921 was the foundation of a joint Faculty of Applied Science and Technology by the Queen's University of Belfast and the Belfast College of Technology, which covered courses in both bodies. Courses were introduced for degrees by evening study alone, including one in naval architecture. Rupert Cameron was one of the first intake of students on this course and he graduated in 1926.

Taking a degree by part-time study while working full time is no soft option and many would comment that it misses out on what is considered one of the main objectives of higher education; that is the broadening experience engendered by mixing with students of different disciplines. To offset this Rupert Cameron believed that *'boys who combined their practical daytime training with their academic study in the evening, emerged fully equipped to take their place in the more responsible posts in industry'*. He also observed that both craftsmen and the staff in his day had surprising interests in a wide diversity of subjects which they pursued in their own time. They were as broadly based in their pursuits as the academics and students in higher education. What is certain is that Rupert Cameron grew up with a broad range of interests in intellectual pursuits.

25,688 ton *Andes*

In 1934 he became the Assistant to the Naval Architect T.C. Tobin who, like many senior staff in H&W, remained in post far beyond normal retirement age. Rupert Cameron was not to become the

Naval Architect of H&W until 1953 when Tobin retired at the age of 77. However, he carried many of the responsibilities of that post from the pre-war years. For example he was proud of his leading role in the design of the 25,688 ton *Andes* delivered in September 1939 and in particular for its *'good lines'*. In 1960 he joined the Board of Directors of H&W.

Naval architecture, like building architecture, involves much more than satisfying the technical specifications such as the hydrodynamic resistance and structural integrity of the hull or the choice and positioning of the propulsion machinery. Just as important are the aesthetics of the design of a ship. It is imperative that a ship should look right as well as satisfying its technical specifications. Rupert Cameron recognised the great value of aesthetics and he deplored the increasing accountant-led approach of ship-owners of tankers and bulk carriers who were uninterested in the good appearance of their ships. Like many great buildings, which can readily be ascribed to a particular eminent architect, it was equally true of the many ships designed by Rupert Cameron.

Successful design not only involves radical lateral thinking but also the ability to assess the consequences and the courage to go out on a limb. Rupert Cameron was a past master of this aspect of design. When Shaw Savill accepted a bid from H&W to design and build a passenger only liner discussions, involving Rupert Cameron, led to the recognition of the potential advantages of positioning the propulsion machinery aft. The enormous advantage was that this would provide a space free of clutter compared with a mid-engined ship but there was much opposition from naval architects and the conservative minded. However, Rupert Cameron could see no valid reasons against it and this led to his design of the first large passenger liner with her machinery aft. The 20,204 ton *Southern Cross* was delivered in February 1955 after having been launched by Queen Elizabeth the ship gave faithful service for forty-nine years.

The success of the *Southern Cross* started a vogue for ships with their machinery aft including the somewhat larger version of the *Southern Cross*, the *North Star*, which was, however, built by Vickers Armstrong (Shipbuilding) Ltd. In particular it was epitomized by the 45,270 ton *Canberra* built for P&O the largest liner built since the *Queen Elizabeth*. This was perhaps the pinnacle of Rupert Cameron's long career. As well as having machinery aft it incorporated 1,100 tons of aluminium in its superstructure, a bow propeller for quick manoeuvring and twin fin stabilisers. The *Canberra* was launched by Dame Pattie Menzies, the wife of the Australian Prime Minister, on the 16 March 1960 and it was delivered on 16 May 1961. After a long and successful career, including serving as a troop-ship in the 1982 Falklands war, it went to the breaker's yard at the end of the twentieth century.

Launching party for the *Canberra*, 1960, R.R.G. Cameron talking to Dame Pattie Menzies

Besides many renowned ships Rupert Cameron was responsible for much else. While the *Myrina* was ordered as a 140,000 ton deadweight tanker, during construction it was increased to 190,000 tons. Rupert Cameron faced the formidable task of precisely sliding the 575 ft of the after end, weighing 17,000 tons, 121 ft 6 in down the slipway to allow an additional length to be added. When the *Myrina* was launched on the 6 September 1967 she was the largest European built tanker of her day. The launch of the three leg oil rig, the *Sea Quest*, on the 7 January 1966 posed a considerable technical problem. It was built with its three legs on separate slipways and despite Japanese computer prognostications of failure Rupert Cameron proved them wrong.

45.270 ton *Canberra* delivered May 1961

Launch of the *Sea Quest* 7 January 1966

In his latter years Rupert Cameron was deeply involved in the detailed specification and layout of the Harland and Wolff building dock which nominally could accommodate a tanker of 1 million tons and its associated new steelworking facilities. This was prompted by enquiries for even larger tankers and bulk carriers and in particular two 250,000 ton tankers. Even while the dock was being built to a very tight schedule the construction of the first of the tankers for Esso commenced. The dock was completed in 1970 when Cameron retired.

The Harland & Wolff Building Dock completed 1970

Shipbuilding is notorious as being a very tough industry frequently run by tough and sometimes ruthless men of iron, such as Sir Frederick Rebbeck. But Rupert Cameron was a very modest, unassuming and friendly individual who had immense loyalty and support from his staff and colleagues. He accepted enormous responsibility while remaining a complete gentleman. From 1970 to 1977 he repaid his debt of gratitude to the Belfast College of Technology by serving as Chairman of the Board of Governors. For his services to industry and naval architecture he was awarded an OBE by the Queen in 1962 and an honorary DSc by the Queen's University of Belfast. He married (1934) Kathleen (d. 1965), daughter of Joseph Lattimore of Carrickfergus; they had one son. He died on the 17 January 1979 at the age of 75.

Sources:
1. 'Shipbuilders to the World - 125 years of Harland and Wolff, Belfast 1961-1986', by Michael Moss and John R. Hume, pub The Blackstaff Press, 1986, ISBN 0 85640 343 1

THE CHAMBERS BROTHERS

Robert Chambers driving a 1905 7hp car in its 3 seater configuration (3rd from right).

The three Chambers brothers were second, third and fifth sons of John and Ellen Chambers of Tullynaskeagh, Downpatrick, Co. Down. They were second, third and sixth of a family comprising seven sons and one daughter. John Chambers was a farmer whose family had been on the same land since 1596. Ellen Chambers was the daughter of Robert Martin, farmer, of Ballywoolen, Rademon, Co. Down, and a first cousin of Sir Samuel Davidson, founder of the Sirocco Works, Belfast.

The brothers were the principals of Chambers Motors Limited, Belfast, the first manufacturer of motor vehicles in Ireland. Between 1904 and 1929 they made a range of private and commercial vehicles most of which incorporated their patented epicyclic gearbox. These vehicles gained a reputation for high quality workmanship and longevity but their unusual design features meant that they also had an unfair reputation of being difficult to maintain and repair.

Robert Martin Chambers, 15 November 1865 - 29 September 1949, was educated at Ballee school and undertook an apprenticeship with the Belfast Shipbuilders, McIlwaine and Lewis. He gained membership of the Institution of Mechanical Engineers through study at the Municipal Technical Institute. Being related to Samuel Davidson, he moved to Davidson & Co., the Sirocco Works as that firm expanded and became a departmental manager. He then moved on to form Chambers & Co., with his brother, Charlie in 1897.

In 1894 Robert married Florence, daughter of Rev. Benjamin Walker of Manchester, and had one son and three daughters. Florence, the eldest, married Professor George Emeleus of QUB. From 1911 he resided in Malone Avenue Belfast. Following the closure of Chambers Motors he designed and developed a range of bakery machinery.

John Henry Chambers (Jack), 29 May 1867 - 29 May 1937, was educated at Ballee school and undertook an apprenticeship with Davidson & Co., Sirocco Works, Belfast. He went to India as a technical representative for that firm whose primary business was the manufacture of drying machinery for the tea industry. Returning to England in 1896 because of ill health resulting from

John Henry Chambers in car entered in the 1906 Scottish reliability trial.

the climate, he became receiver and manager of James Wilson & Co., Vauxhall Iron Works, London. This firm, which made internal combustion engines for river launches, required to diversify and chose to manufacture motor cars. Jack was a co-designer of the first Vauxhall car in 1903. He left the firm in 1906 when it moved to Luton and spent a year as a consultant in London. He joined his brothers in Chambers Motors Ltd., in 1907.

In 1897 Jack married May Cranbrook of London, who died in 1906, and had four daughters. He later married Margaret Benett but had no further family. He and his brother Charlie drove Chambers cars in Irish and Scottish Reliability Trials in 1906, 1907 and 1908 gaining several class wins. He was President of the Belfast Association of Engineers in 1918 - 9. He resided in Sandown Road, Belfast.

Charles Edward Chambers, 21 June 1873 - 6 September 1931, was educated at Ballee school, and the Royal Belfast Academical Institution. He undertook an apprenticeship with, and then gained employment as a draughtsman in the firm of Harland & Wolff. He left to form Chambers & Co., with his brother, Robert, in 1897.

Charlie married Elsie Holland and had one son and one daughter. He lived in Kings Road, Knock, Belfast. He was responsible for the finances of Chambers Motors Ltd. and the stress of liquidation is reported to have caused his early death.

In 1897, Robert and Charlie founded Chambers and Company, in Cuba Street, Belfast, as a general engineering business to provide support to the many larger firms then active in Belfast. They designed, developed and manufactured a machine which wired the corks into bottles of aerated waters, the product of another Belfast Firm, Grattans of Cromac Square. Four patents taken out by Robert and Charlie, viz., 13803 of 1898, 6389 of 1899, 13446 of 1899 and 15322 of 1903 detailed their ideas and improvements in and relating to these machines. Their product was not only used locally but was also exported world-wide. They also designed, patented and built an overhead travelling crane for handling carcasses in the Belfast Abattoir. This was detailed in patent number 16435 of 1900.

Meanwhile, in London, their brother Jack as managing director of J. Wilson & Co. Ltd., the Vauxhall Iron Works, was working on light, internal combustion engines for marine applications. He took out patent 26085 of 1901 "Improvements in Bearing for Rocking Shafts

Charlie Chambers at the wheel of his car in Craigantlet hill climb in 1912. The Dunlop Company used this photograph to demonstrate the strength of their inner tubes.

and the like." When he started designing the Vauxhall car he developed an epicyclic gearbox which he patent under 28577 of 1903 "Improvements in Change Speed Driving Gear for Motor Cars."

He sent his brothers an example of the first car for their evaluation and comments. As they were driving from Belfast to the family home at Tullynaskeagh, it broke down and they were forced to stay overnight in the Baloo Inn. They decided they could make improvements so they built their own car. They were so proud of their handiwork that they sent a telegram back to the works to announce their safe arrival in Bangor (12 miles from Belfast) on the first run at Easter 1904. As a result they started manufacturing a twin-cylinder, horizontal engined car, the first to be made in Ireland. The whole design can be seen in their patent, number 11351 of 1904. This car was rated at 7hp and had some unusual features including the epicyclic gearbox which was arranged to provide three forward speeds and reverse from a single sun, planet and annulus gear set, an improvement of Jack's design. Had they been working in one of the areas normally associated with the motor industry, conventional wisdom would have told them it was impossible. However, as no one advised them not to, they went ahead and made a success of the idea. The crankcase had a large inspection door through which the pistons could be removed and the bearings inspected without disturbing any other component. This was a successful idea which was taken from marine engine practice. The drive was taken from the engine to the rear axle by a large chain. This was arranged in a chain case which acted as an oil bath and which could also be adjusted to maintain a constant tension and allow for wear and stretching of the chain. Other features included a steering wheel which could be hinged up to give better access to the driver's seat and this seat together with the passenger one could be swivelled out to allow a third or "Shamrock" seat to be installed.

The car proved to be quite popular and the business expanded. Many early clients appear to have been Doctors and several different styles of bodywork became available including closed versions to protect the occupants from inclement weather. Such bodies, being heavier, required a longer chassis and a more powerful engine. A series of increasingly powerful cars retaining the twin cylinder design was produced. Jack joined his two brothers in 1907 and the business became a limited liability company styled Chambers Motors Limited. The brothers continued to develop the design of their products and took out two more patents, 26422 of 1906 for "Improvements in Power Transmission Gear for Automobiles" and 7747 of 1908 "Improvements in High Tension Magneto Ignition". The firm used the popular motoring events of the period such as the Irish and Scottish Reliability Trials between 1906 and 1908 to publicise the cars. All the brothers competed but Jack was the more successful driver, gaining several silver medals for his efforts.

Competition experience showed that their design was in need of updating. In 1909 they introduced a four-cylinder model again with a transverse, horizontal engine. The prototype proved very successful but the production models suffered from repeated crankshaft breakages. After investigation, it was discovered that the material supplied for the crankshaft of the prototype had been incorrect and of a stronger and more resilient composition and was much more expensive. However, due to the rapid improvements in design and reliability of their competitors, the brothers decided to adopt a conventional, longitudinal, vertical four cylinder engine for their new cars from 1911 on. They continued to use their patented epicyclic gearbox, still located in the rear axle casing but now with a shaft drive. This design called for a large diameter ball bearing which had to be bought from Germany. With the onset of the First

World War, the supplies of this component ceased and Robert Chambers successfully devised a way to manufacture it on his own limited machinery.

The brothers prided themselves on the fact that the firm designed and made all the components themselves. Production methods involved a mixture of craftsmanship and brute force. Jigs, which they made themselves, were designed for batch production but batches were in single figures rather than the hundreds which could have been accommodated. Rear axle casings were produced in jigs which could be accommodated on a turret lathe to save the expense of buying a dedicated machine tool. Engine casings and chassis side members were individually marked off from blueprints then drilled and tapped by hand as necessary. The propellor shaft cover, an interference fit into the rear axle housing, was driven home by a fitter with a sledge hammer.

During 1910, in response to increased demand, a new and separate body shop and paint shop had to be opened on the Newtownards Road near the main factory which was situated in Cuba Street. In 1913 the whole operation was moved to larger premises across town in University Street, where all the activities could be carried out under one roof. Showrooms were opened in Chichester Street, then the centre of the Belfast motor trade. Two four cylinder models called the 11/15 [Bore 3 1/8 ins, (80mm), Stroke 4ins, (102mm), - 2052cc] and the 12/16 [Bore 3 3/8 ins, (86mm), Stroke 4ins, (102mm), - 2370cc] were available and a large variety of bodies could be supplied for either private or commercial use. At that point, the firm was at its most successful.

Production of private vehicles came to a halt during the First World War due to lack of components as engine castings, for example, could not be purchased from the supplier in Coventry. However, a few cars were completed from stocks of components. War Office requirements of a minimum engine bore of 100 mm (as opposed to their 86 mm) prevented the company from competing for official contracts for lorries or ambulances. However, a fleet of six ambulances, ordered and paid for by local public subscription, was built. These were driven and maintained by volunteers and served with the Ulster Division in France. At the same time, the company manufactured 18-lb shell cases and hand grenade percussion caps in large quantities. They used their existing equipment and machine tools with a series of jigs specially designed for each operation. One special purpose machine, which Robert Chambers devised, was capable of turning out 7,000 finished percussion caps in 24 hours. He received a Ministry of Munitions reward of £150 for the design. One of the differences between working in Ireland and the rest of the United Kingdom was that there was no new machinery available for munitions work so the existing machinery became worn out more rapidly than normal. With the return to peacetime conditions the majority of the machinery had to be replaced whereas the competition in England had been re-tooling with suitable machinery which was still viable for making automobile components.

1920 Open Drive Saloon

Post - war production started in 1919 with up-dated versions of the pre-war designs. To keep working during the material shortages experience as a result of the many strikes in England, the firm bought in some of their older cars and remanufactured them as commercial vehicles and became agents for Karrier and Renault commercial vehicles. Robert kept developing and patenting new ideas and for components. Patents 163905, "Improvements relating to Vehicle Hoods", 172159 "Improvements in Stops for Doors, windows etc.," 176602, "Improvements in Sliding Glass Windows", and 181583, "Improvements in Spring Catches or Fasteners", were all granted in this period. From 1925 they also produced a few examples of two new, and more up to date, designs buying in proprietary components such as engines, gearboxes and axles. They maintained their high standards of quality and workmanship but, in the face of increasing competition from English and American firms which adopted mass production, Chambers Motors Ltd. went into voluntary liquidation in 1929.

After the closure of the company, Robert used part of the premises to set up business as a manufacturer of machinery for the bakery industry and took out patent number 293966 "Improvements relating to the Rolling and Pinning of Dough.". Until his early death, Charlie and his son Billy operated a car repair business in another section and the remainder was leased to the Belfast Omnibus Company. Jack continued to offer his services to the motor industry as a consultant in gearbox design, taking out patents 161613, 315476, 315920 and 337744, all of which related to improvements in variable speed epicyclic gearing.

Sources:
1. 'Motor Makers in Ireland', John S. Moore, pub. The Blackstaff Press, Belfast, (1982) ISBN 0 85640 246

2. 'Chambers Motors 1904 – 1929', John S. Moore, pub. Dreoilín Publications, Tankardstown, (2000). ISBN 1 902773 09 8

WILLIAM DARGAN

William Dargan was born on 28 February 1789 son of a farmer in County Carlow. He was educated locally and undertook his initial training as a surveyor. He was introduced to Thomas Telford and worked for him on the reconstruction of the London to Holyhead road when it was being upgraded as a primary mail route. The stretch along the coast and the causeway at Holy Island has been attributed to him. He then returned to Ireland to set up on his own and became a contractor for a variety of civil engineering projects initially concentrating on roads. The Dublin to Howth road was the most notable of this period.

In 1833 and 1834 Dargan built the first railway in Ireland, the Dublin to Kingstown, where his experience of constructing embankments on soft foundations must have been essential. His obituary in the Belfast Newsletter of 8 February 1867 contains an insight to the thinking of the time. 'It is related that at a soiree given one night at a distinguished house in Dublin, railways became a topic of conversation and a person present suggested a line between Dublin and Kingstown. Very good: but where was the money to be got? What would it cost? One sum was named - another was hazarded. But what a difference between them. The idea was about to die out in a laugh, when the first speaker said "Here's who will tell us in a moment. Here, Dargan, yours is the head for calculations, what would a line of rails to Kingstown cost?"

Tablets were out and a pencil writing down a few figures. In two minutes a result was announced - so low as to astonish everyone present; and it was agreed to meet the next day and consider the project. The company was formed, the Act of Parliament obtained and in due time tenders for the contract were invited. It was the first bit of railway in Ireland. Most of the tenders were ridiculously high; but Mr Dargan sent in the same rough draft that he had exhibited at the soiree and got the contract.'

He went on to build many other railways throughout Ireland and, in total, completed some eight hundred and sixty miles of lines out of a total of eleven hundred at the end of 1860. Due to the problems of financing the large number of railway companies in such a relatively small island, he often had to accept payment in the form of shares in the company. This practice resulted in him holding directorships in several railway companies, but only when he was no longer under contract to them. He became chairman of the Dublin, Wicklow and Wexford Railway.

The compiled list of his railway activities is most impressive.

D&KR	Dublin Kingstown	1834	6 miles
UR	Belfast Lisburn	1839	7 miles
UR	Lisburn – Portadown	1842	$17^{1}/_{2}$ miles
D&DR	Portmarnock – Drogheda	1844	$24^{3}/_{4}$ miles
GSWR	Dublin (Kingsbridge) – Carlow	1846	56 miles
D&DR	Howth Branch	1847	$3^{3}/_{4}$ miles
GSWR	Mountrath – Ballybrophy	1847	$7^{1}/_{4}$ miles
GSWR	Carlow – Bagnelstown	1848	10 miles
GSWR	Ballybrophy – Thurles	1848	21 miles
B&HR	Belfast – Holywood	1848	$4^{1}/_{2}$ miles
B&BR	Belfast – Carrickfergus	1848	$9^{1}/_{2}$ miles
B&BR	Belfast – Ballymena	1848	$30^{1}/_{2}$ miles
UR	Portadown – Armagh	1848	$9^{3}/_{4}$ miles
W&LR	Limerick – Tipperary	1848	$24^{3}/_{4}$ miles
GSWR	Thurles – Cork	1849	78 miles
D&ER	Dundalk – Castleblney	1849	22 miles
N&WR	Newry – Warrenpoint	1849	6 miles
CB&PR	Cork – Passage	1850	$6^{1}/_{4}$ miles
ISER	Bagnelstown – Kilkenny	1850	16 miles
B&CDR	Belfast – Newtownards	1850	$12^{1}/_{4}$ miles
MGWR	Mullingar – Galway	1851	$27^{1}/_{4}$ miles
B&DJR	Portadown – Mullaglass	1852	$16^{1}/_{2}$ miles
W&LR	Tipperary – Clonmel	1852	$22^{3}/_{4}$ miles
W&TR	Waterford – Tramore	1853	$7^{1}/_{4}$ miles

W&LR	Clonmel – Waterford	1853/4	27¾ miles
DWWR	Dublin (Harcourt St.) – Bray	1854	14¾ miles
GSWR	Portarlington – Tullamore	1854	15½ miles
KJR	Mallow – Killarney	1854	40 miles
BB&CJR	Ballymena – Coleraine	1855	28¾ miles
BB&CJR	Coleraine – Portrush	1855	5¾ miles
DWWR	Bray –Wicklow	1855	14¾ miles
MGWR	Mullingar – Longford	1855	26 miles
MGWR	Inny Jct. – Cavan	1856	24¾ miles
B&NCR	Randalstown – Cookstown	1857	27 miles
L&FR	Limerick – Foynes	1858	26¼ miles
UR	Armagh – Monaghan	1858	16 miles
P&DR	Portadown – Dunganon	1858	14 miles
L&ER	Limerick – Ennis	1859	24¼ miles
T&KR	Killarney – Tralee	1859	61½ miles
BJR	Scarva – Banbridge	1859	6¾ miles
GSWR	Tullamore – Athlone	1859	23¼ miles
A&TR	Athenry – Tuam	1860	15½ miles
M&FR	Mallow – Fermoy	1860	17 miles

Dargan's Railways

As well as his work on the railways, Dargan built many roads and canals throughout Ireland. His speciality seems to have been in areas with drainage problems. Many of his railway embankments were constructed to reclaim large areas of tidal marshland and still provide the coastal defences. In 1841, after years of hesitation and enquiries, the Harbour Corporation, which was responsible for the Belfast Docks, straightened part of the River Lagan and created new quays between Dunbar Dock and the First Bend. The work was carried out by William Dargan. The spoil from this First Cut was deposited on the east bank of the cut to form Dargan's Island. As a result of the Belfast Harbour Act of 1847, the newly formed Belfast Harbour Commissioners were able to dig a Second Cut to give a straight channel from the sea to Belfast Docks. This work was also undertaken by Dargan but given the name Victoria Channel when it was opened in July 1849 in honour of the Queen's visit to Belfast the next month. Dargan's Island, originally a pleasure garden, subsequently became Queen's Island and the home of the shipyards.

During the famine years of 1845 - 47 he retained those of his men for whom there was no work to help alleviate the hardship unemployment would cause.

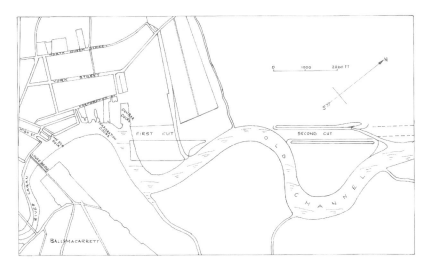

Plan of Belfast Harbour 1843

Dargan provided financial backing for the Great Exhibition of 1853 in buildings on Leinster Lawn, Dublin, which became the National Museum. He placed some £30,000 in the hands of the organising committee to initiate the project. By the time of the opening on 12 May 1853 his guarantee against loss stood at nearly £100,000 and by the final reckoning, he is believed to have lost a total of £20,000. Queen Victoria paid him the compliment of visiting his home, Mount Annville and offering him a baronetcy, which he declined.

He endeavoured to encourage the development of flax growing in the south and east of Ireland, including the running of a couple of mills. However, this project was mismanaged for him and he lost a part of his fortune. He served on the Dublin Corporation for some time and was elected High Sheriff of Dublin in 1863.

He was seriously injured in a fall from his horse in 1866. This incapacitated him quite seriously and his affairs were supervised most inefficiently, which in turn affected his health and will to live. He died at 2 Fitzwilliam Square, Dublin on 7 February 1867 and was buried in Glasnevin Cemetery. His widow, Jane, was granted a civil list pension of £100.

His place in the story of the railways in Ireland is remembered with the new bridge built over the River Lagan in Belfast in 1990, to join the former B&NCR and GNR(I) sections of Northern Ireland Railways, being named in his honour.

Sir SAMUEL DAVIDSON

Samuel Cleland Davidson was born on 18 November 1846 in a house in Bridge End, Ballymacarrett, Belfast. He was the youngest of a family of seven, five boys and two girls. Their father, James Davidson, owned a flour mill in Bridge End and their mother was Mary Taylor of Ballymacrae, Co. Down.

Davidson was educated at the Royal Belfast Academical Institution, leaving school at the age of fifteen. In later life he claimed that the most important thing he learned was to be observant. In the following two years besides being tutored in various subjects he served for a period in a surveyor's office and then 'doing the books' in his father's flour mill. If this was planned experience is not known, but what is certain is that it stood him in good stead for what was to come. In 1864 he joined his cousin, James Davidson, an early pioneer tea planter with the Assam Company, to help him run some tea estates in which the Davidson family had an interest.

Davidson left Belfast for Calcutta in August 1864 to travel to the Cachar District of Assam, which is a remote State of India in the North East adjoining Bhutan and Tibet. It involved travelling 600 miles overland from Calcutta mainly by boat under primitive conditions, arriving in December 1864 shortly after his eighteenth birthday.

In 1864 Assam was a tough and unhealthy environment but Davidson studied preventive medicine and he survived and flourished, while many of his fellow countrymen died or suffered ill health. Within a few months it was evident that he was an outstanding character. He modernised book keeping and drew up new maps of the estates, he designed himself a type of bungalow, which was widely adopted and, more importantly, he developed the cultivation and manufacture of tea. By careful recording, analysis and experiment he established the factors to produce good quality tea whereas before it had been a matter of chance.

The Tea Industry in Assam

After two or three years, when he was in his early twenties, he was promoted to take over the management of Burkhola, a larger estate, and his expertise was widely sought after. However he was dissatisfied with the crude and inefficient methods of withering – a fermentation/oxidation process – rolling the tea and drying it before finally sorting and sizing the tea. He developed a cylindrical drying machine, which he patented in 1869, and a year later he obtained a patent for a tea roller. This was the start of his life long interest in tea making machinery.

In 1870, after returning to Cachar following his first home leave, he purchased his own tea estate and increasingly devoted his attention to the tea drying process, which he recognised as pivotal to producing good quality tea. He returned home again in 1872 to marry Clara Mary Coleman of Belfast who accompanied him back to his Subony Estate where their first daughter, Annie, was born in November 1873. When Annie died the following year Clara Mary returned to Belfast in the interest of the health of their future children.

Davidson continued with his own hands and the assistance of local labour to construct one drying machine after another until, in 1875, he had developed a satisfactory machine, which he patented. But by that time he had recognised that it needed to be a high class engineering job so he returned to Belfast where he got his friends in Ritchie Hart and Co to construct his latest design. On his return to India, in 1878, he demonstrated the machine widely throughout the country and obtained substantial orders, while at the same time selling a simple type of wheel hoe which he had patented.

On his return to Belfast the earliest dryers were produced for him by Ritchie Hart and Co. but in 1881 he set up his own engineering company in Bridge End employing a handful of workers. The success of his first dryer was immediate and it led to a never ending development of both the static up-draught and down-draught dryer. These depended on the up-draught induced by the furnace chimney. This was a serious limitation and to improve the efficiency of drying a positive pressure was needed. At that time, however, the only fans available were hopelessly inefficient.

Davidson's first factory in Belfast.

The Factory in 1898

At that period mechanics of fluids was an ill understood science and there was no recognised method of testing fans nor were the scientific instruments to measure air flow available. Davidson proceeded to construct a very simple but elegant rig to measure the relative effectiveness of fans. In this rig the air was blown down a square sectioned duct and at the far end there was a heavy wooden door hinged at the top with a quadrant to measure the angle the door was forced open. This rig allowed him to place different designs of fans in an order of merit.

Fan Test Rig

With this test rig Davidson proceeded to test a whole range of fans of different geometries. In 1898 one of these gave a wholly unexpected result, when it blew the door through 90 degrees. This was the birth of what became known as the 'Sirocco Forward Bladed Centrifugal Fan', on which the future of the company was built. Not only was it applied to designing ever more compact tea dryers, but it had wide application for ventilation systems, forced draught boilers, mine ventilation and much else.

It was during the course of developing fans that a planter friend visited the works. Astonished at the volume of air being produced he exclaimed "Why it's just like the Sirocco wind that blows off the desert". Davidson, ever a man with an eye to publicity, adopted the name as his trade name.

Less well known is that Davidson devised the highly innovative scheme for the heating and ventilation of the Royal Victoria Hospital in Belfast opened in 1903. This, for the first time, incorporated an elementary form of environmental control. Davidson was not only responsible for its design but also its construction, installation and maintenance. His concept had a major influence on the architectural design of the building.

Two axial fans, driven by steam engines operated from waste steam from the boilers of the hospital's laundry, delivered air to an immense brick lined duct, which ran the length of the hospital at ground level. The wards, one floor up, were at right angles to the duct. Distribution channels from the main duct fed air through risers which delivered air through openings above head level in the wards. Foul air was dispersed through slots in the skirting around the perimeter of the rooms, into an exhaust duct system.

The heavily polluted air of Belfast at the beginning of the twentieth century was drawn into two heating chambers, through hanging curtains of coconut fibre, kept moist by sprinklers in the roof of the chamber. In winter the water for the sprinklers was heated to prevent freezing. The cleaned air was then passed through heating coils and extra heating, if needed, was provided at the entrance of the distribution channels to the wards.

In the winter it was recognised that the cold air passing through the coconut curtains had, after heating to the sixties Fahrenheit, a relatively low humidity. This was corrected by water sprays adjusted to the individual need by the humidity recorded in the wards by wet and dry bulb thermometers. This was a most innovative concept at that time.

It is only within the last few years, nearly a century after the RVH was opened, that it has been replaced by a modern building.

During the 1914-18 war he turned his mind to helping the allied cause. He provided over 8000 fans to the Royal and Merchant Navies, and it was ironic that many of the ships of the Kaiser's fleet scuttled at Scapa Flow were equipped with Sirocco fans. Towards the end of the war Davidson had developed a 40 mm hand held mortar which the Americans were getting tooled up to mass produce when the war ended.

In 1921 Samuel Davidson with over 120 patents to his name received the KBE from His Majesty King George V and he died two months later. His wife predeceased him in 1918. Their lives had been rather tragic as not only did their first daughter die when she was eight months old but their third child, James Samuel, born in 1877 was killed on the Somme in 1918 and their fourth child, Richard Frederick, born in 1878, died of meningitis in 1897. Two daughters Clara May, born in 1875, and Kathleen, born in 1882, survived. The son of Clara May, Edward D. McGuire, ultimately became Chairman of Davidson and Co Ltd.

After Davidson's death the company continued to prosper and expand under his impetus into producing ever larger fans including very large axial flow fans and into dust control plant. After the Second World War they continued to retain a large fraction of the market for tea making machinery as well as fans for all kinds of application. They also were involved in gas circulators for the gas cooled nuclear reactors and very large re-generative air pre-heaters.

In the 1970's the Company suffered from a serious decline in the market for their equipment. After many vicissitudes they were bought by their main competitor, in 1988, to form Howden Sirocco. This company still exists in Scotland and Belfast, but with a greatly reduced workforce, and they have set up their main manufacturing plant in Weihai in China, which was opened in 1996.

Howden HUA Engineering Company Ltd. (China).

Sources:

1. 'The Sirocco Story' by Edward D. McGuire, pub. Davidson and Co Ltd, 1969.

2. 'The Man of a Hundred Inventions: Founder of Davidson and Co Ltd', pub The Times of Ceylon, Jan 28 1937.

3. 'The Architecture of the Well-tempered Environment' by Reyner Banham, pub. University of Chicago Press, 1969, ISBN 0226036979

4. 'Engineering! Ireland v China' by W.B.A. Stafford to be published in The Engineers Journal of the Institution of Engineers of Ireland.

Sponsored by Howden Power Limited

JOHN BOYD DUNLOP

John Boyd Dunlop was born in Dreghorn, Ayrshire on 5 February 1840 son of John Dunlop and his wife, Agnes Boyd. The family had been farmers in the area for many generations but his father seems to have been an innkeeper.

Because his health did not seem robust enough for him to become a farmer, Dunlop was educated at the Dreghorn Parish School and Irvine Academy before going to the Royal Dick Veterinary College in Edinburgh from where he graduated as a veterinary surgeon at the age of nineteen. He set up his plate in Belfast about 1867 and soon developed a major practice in the centre of town adjacent to the markets and carriage builders. In December 1871 he married Margaret Stevenson the daughter of a Ballymena farmer. They had two children, John Boyd junior (Johnnie) and Jean.

At this period, Belfast was well provided with streets of durable cobbles and granite square setts. Any horse drawn carriage travelling along these streets had the wheel vibrations damped out by the long supple springs on which it rode. However, the dashing young men about town wanted to show their own prowess and partake of the craze of cycling which rapidly gained popularity in the eighteen eighties. In cycling, where the wheel had only a thin rubber tyre, these vibrations, known as "jarring", passed straight from the wheels to the rider's body. Young Johnnie was as keen as any other youngster to join in the craze and the family physician believed that the exercise would be good for him if it were not for the jarring.

The Dunlop Family

About this time, Dunlop, who had a long interest in the problems of transport and had looked at the possibility of spring wheels, was toying with the idea of a pneumatic cushion in a horse collar to even out the pressure on the horse's neck and shoulders when under strain. He had discussed the principle with Sir John Fagan, founder of the Belfast Hospital for Sick Children, who had related

Reconstruction of the initial Dunlop experiment with the solid disc wheel.

his experiences of the benefits of air mattresses as opposed to water filled ones. The similarity of the problem with jarring of cycle wheels suddenly became clear and the question of how to apply a pneumatic cushion to a wheel was addressed. Dunlop adopted an experimental approach to the problem and took a wooden disc to which he added an air tube round the circumference. He held this in place with a strip of canvas nailed to the disc. When one with the air tube inflated was rolled up the stable yard it went much further than a conventional, spoked tricycle wheel of the same diameter.

The next step was to construct special tyres for Johnnie's tricycle. These were made of rubber tube which was held in place on the wheel rims with specially chosen linen and had an extra layer of rubber in contact with the road. The whole tyre was held in place by having the linen soaked in rubber solution and wrapped round the inner tube and the wheel rim. This construction, later called the "mummy" tyre because of its similarity to the Egyptian embalming technique, proved very successful and Johnnie was able to beat all his school-mates when racing over both cobbles and on prepared cinder tracks.

Dunlop patented his idea in July 1888 and soon other people started asking him to make copies of the tricycle for them. As his main interest was in his large and thriving practice, he allowed Robert Edlin and Finlay Sinclair, Belfast cycle makers, to make some bicycles and tricycles which could use the new tyres. Because the design of these new pneumatic tyres were so much wider than the conventional solid ones, new forks and brackets had to be cast for these machines. Edlin and Sinclair had also to be instructed in the secrets of tyre making. The "mummy" was normally only applied to safety bicycles but the Ulster Folk and Transport Museum has a unique high wheeler, better known as a penny farthing, with an original "mummy" tyre in its collection.

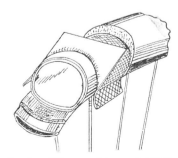

Dunlop "Mummy" tyre of 1888

Cycle racing in the 1880's was very much the premier sport for young men. The ordinary or "penny-farthing" cycle with its large front wheel was just the job and gave some spectacular sport. However the rider sat high above ground level and could easily go over the handle bars in an "imperial crowner" which was frequently fatal. There was a section of the racing fraternity who believed that the "safety", with its equal sized wheels, was better but the exponents of the ordinary called them "cissies" and often did not allow them to race. Willie Hume, captain of the Belfast Cruisers Cycle Club, had taken an imperial crowner and had considered giving up racing in which he had been moderately successful. However, having had an opportunity to try a pneumatic tyred bicycle, he to keep racing on it. He entered for the Queen's College sports on 29 June 1889. In this match he beat the reigning Irish champions Arthur du Cros and R. J. Mecredy. To prove that his win, first time out, was not a flash in the pan he followed it up with a win in a 17 mile road race from Belfast to Antrim the next week-end.

Du Cros and Mecredy were rapidly converted to the new design and won many races during the hectic summer of 1899. It came to the point where their entries to English race meetings were being refused to protect the reputations of the top English riders. It rapidly became evident to those who were not making the bicycles that Edlin and Sinclair could not keep up with the demand for both bicycles and tyres. A Dublin cycle agent and financier, William Bowden and his partner J. M. Gillies, therefore approached Dunlop with the object of floating a company for the purpose of increasing the availability of the pneumatic tyre. Bowden and Gillies then approached a Dublin paper merchant, Harvey du Cros, President of the Irish Cyclists Association and father of the Irish Champion, Arthur, to help the in launching the company. The condition which he imposed was that he should have complete control, appoint directors, write the prospectus and make the issue to the public. The firm which introduced the pneumatic tyre industry to the world was founded as the Pneumatic Tyre and Booths Cycle Agency in Dublin in 1889. By incorporating the Booth's cycle agency and Edlin's business in Belfast, the firm had a fall-back position should the tyre fail to make the anticipated profits. Edlin, who only had a verbal agreement with Dunlop, had to join the company as an employee. He received only £490 in shares as payment for his earlier services. Edlin left at an early stage but his assistant, Sinclair, stayed and rose to become works manager of a much expanded operation.

When interviewed at the time, Dunlop expressed the opinion that the patent was valid and there was a commercial future for the tyre on bicycles, for invalid carriages and all horse drawn vehicles. The sporting success of the pneumatic tyre meant that many other inventors were trying to find ways round the patents and unscrupulous manufacturers were breaching the patents by simply making tyres without a licence. However, in the autumn of 1890 it was suddenly discovered that Dunlop did not hold a valid patent as the principle of a pneumatic tyre had been described in a patent of 1845. Then, another Scot, William Thompson, had designed "Aerial Tyres" which used leather tubes to contain the air. This material was not sufficiently airtight to use. The idea had not been a success and had been forgotten.

Dunlop was prepared to abandon the pneumatic tyre and concentrate on his later ideas of a spring-frame bicycle and cushion tyres. Harvey du Cros thought differently. He decided his investment was worth fighting for and put up a spirited defence. He acknowledged the impracticalities of the non-detachable tyre and with an employee, a Belfast cycle racer Thomas Robertson, came up with a detachable cover. This used two circumferential lengths of wire which were fed through the rim and secured with nuts. Unfortunately for the company they had again been anticipated. This time Charles K. Welch had patented the now universally adopted "well base" tyre and rim. Du Cros bought the patent from Welch and employed him as a technical adviser to the company. From 1893, the tyres were sold as Dunlop-Welch detachable tyres.

Another important patent was bought about this time. Charles Woods, whose brother Frederick was one of the directors of the Pneumatic Tyre Company, designed the valve which rapidly became the universal cycle tyre valve. It had a rubber sleeve which acted as a one way valve on a metal finger that had a simple drilled passage down the centre and emerged radially. He asked £1,000 for his patent and refused a choice of either 3d per unit as a royalty payment or 11d a unit if he were to supply them. Either alternative would have suited the cash strapped company but Woods wanted to remain in his cotton spinning business. By holding out for the cash payment, he lost out on a much bigger fortune.

Woods Valve Patent No. 4175 of 1891

While this was happening in the summer of 1890, the Pneumatic Tyre Company had opened a branch in Coventry, the centre of the Cycle Industry, to reduce the expense for smaller cycle manufacturers in having to send wheels and men to Dublin. Then in the spring of 1891, the Dublin Corporation, supported by the Dublin Trades Council, took an action against the company on the grounds of nuisance created by the smell of rubber solution. They asked for the removal of the business. Although the Corporation lost its case in the lower court, it served notice of appeal. The directors decided that enough was enough and, as the major part of their business was then with English cycle manufacturers, they would move to Coventry.

While the Pneumatic Tyre Company sold detachable tyres from 1893, they had to fight a series of legal battles against patent infringements. It was not until 1895 that the House of Lords ruled that the Welch parent was the master patent for wired on tyres. The main competitor, the Clincher Tyre, which was held in place by internal air pressure, had been patented in 1890 by William Bartlett of the North British Rubber company. Made under licence by the Michelin brothers in France, clincher tyres were demonstrated by Emile Levassor on his car in the Paris - Bordeaux - Paris race of 1895. No case had been taken to define the relationship between the two master patents or to see if one design could dominate the market. Counsel were being briefed for what looked like being a long and costly battle when du Cros met Bartlett and offered to buy his patent while allowing NBR to continue making tyres and receive the royalties from Michelin. This resulted in the formation of the Dunlop Rubber Company. The shares of the Pneumatic Tyre Company, which had a face value of £1, were bought for £12.10 0 each giving the business a value of £3,000,000 compared with the stock market value of £1.4 million. The new company was floated with a value of £6,000,000 which left a considerable profit for du Cros and the other financial backers. With du Cros as chairman, the Dunlop Rubber company became a world leader for over fifty years.

The Pneumatic tyre company itself had been a most profitable venture. R. J. Mecredy, a director, calculated that if a man invested of £100 in the original shares in 1889 and took up the proportion of shares allotted to him in subsequent issues by 1896 held 1,038 shares valued at £12,658 for which he had paid £1,146. He had also received over £1,700 in dividends.

Dunlop severed his relations with du Cros in 1894 when he sold his shares in the Pneumatic Tyre Company and resigned as a Director.

Dunlop leading a gathering of old time cyclists in 1917

He bought a controlling interest in Todd, Burns & Co., a Dublin department store. He also sold his veterinary practice and invested in other commercial ventures thus becoming able to lead the life of a gentleman.

The company had always used a portrait of Dunlop with his instantly recognisable flowing beard as a trade mark. Shortly before his death, he took objection to the addition of a smartly dressed gentleman. He said the wearing of spats made him out to be a dandy rather than a hard working vet.

Dunlop died at home in Dublin on 23 October 1921.

Sources:

1 'The History of the Pneumatic Tyre', by J. B. Dunlop, pub. Alex Thom & Co. Ltd., Dublin 1924

2 'Wheels of Fortune', by Sir Arthur du Cros, pub. Chapman Hall, London, 1938.

3 'Motor Makers in Ireland', by John S. Moore, pub. The Blackstaff Press, Belfast, 1982. ISBN 0 85640 264 8

HARRY FERGUSON

Henry George Ferguson, better known as Harry Ferguson, was born on 4 November 1884 on the family farm at Growell near Dromara in County Down. He was the fourth child in a family of eleven dominated by a tough, uncompromising and God fearing father, James Ferguson. His Scottish ancestors migrated from Scotland to the Plantation of Ulster in the early seventeenth century. Harry's mother, Mary (neé Bell), who came from Newry was gentle and loving. Her two step sisters became the first women doctors in Ireland. He attended two local schools and finished his education at the age of fourteen and started work on the family farm.

1862 Burrell-Fowler Ploughing Engine.

In the mid nineteenth century stationary and portable steam engines became available for driving agricultural machinery, such as threshing machines. Then in the eighteen fifties John Fowler, inspired by the tragedy of the Irish famine, designed and developed the steam plough and set up the Plough Works in Leeds to manufacture them. This involved two steam traction engines fitted with powered cable drums, one each side of the field, which pulled a double headed plough to and fro across the field, while the engines slowly inched forward. For the rest of the century steam ploughs provided the only practical, mechanical, plough and were extensively adopted world wide, but only on large farms and estates. The smaller farms, which were responsible for the bulk of farm output, could not afford mechanisation and they continued to rely on horse drawn farm implements.

Harry Ferguson, spurred on by his tough experience on his father's farm, was to be instrumental in revolutionizing the farming industry by what he called the 'Ferguson System'. His contribution has proved to be the most important development in agriculture in the twentieth century. It has made it possible to feed many of the world's burgeoning population.

Ferguson, who was of light build, was not physically well suited for farm labouring and he was diametrically opposed to the tough regime and beliefs of his father. In 1902 he had decided to immigrate to the New World when his elder brother, Joe, offered him an apprenticeship. Joe had left home in 1895 to take up an apprenticeship as a maintenance mechanic in the Combe-Barbour Mill Engineers. In 1901, with another ex-apprentice, he set up Hamilton and Ferguson, an engineering business. After Harry joined them, Joe broke up the partnership and in 1903 set up J. B. Ferguson & Co., Automobile Engineers,

Like his brother Harry was immensely interested in cars and motor cycles and particularly in internal combustion engines, as were many young men at that time. Engine tuning was an art rather than a science and he became recognised as a gifted engine tuner. This contributed to the increasing prosperity of the company. At that time he attended evening classes in the Municipal Technical Institute, which was being rapidly developed under the Principal, Mr F.C. Forth. Like the vast majority of engineers at that time this was to be the only technical education he was to receive, but clearly it stood him in good stead.

In 1908 he became interested in the possibility of designing and manufacturing aeroplanes, which had excited the imagination of the public since the first powered flight by the Wright brothers in December 1903. He attended flying displays in England and Europe to study the design of aeroplanes on display and note their dimensions. When he decided to build his own aeroplane he adopted the tractor monoplane configuration used by Blériot in his successful first flight across the English Channel on 25 July 1909. In his final design Ferguson used a 35 hp JAP engine. After experimenting with various designs of propellers he first successfully flew a short distance on 31 December 1909 at Hillsborough Park, the estate of Lord Downshire. This was the first flight in Ireland and Ferguson became the first Briton to build and fly his own monoplane. It was an example of his dogged perseverance which Ferguson demonstrated throughout his life.

Ferguon flying his reconstructed monoplane in October 1911 at Newtownards – note tricycle undercarriage.

In April 1910 Ferguson managed a flight of a mile at Massarene Park, Co. Antrim. In June he made several flights at Magilligan Strand, Co. Londonderry, including one of 2½ miles. After problems and accidents he managed to fly over 3 miles on 5 August 1910 at Newcastle, Co Down, to win a prize of £100. Subsequently he made several more flights at Magilligan Strand including carrying passengers. In October he was knocked unconscious in a crash which completely wrecked his aeroplane. However he did not give up, but preceded to radically redesign the fuselage and improve the reliability and the form of the engine. In June 1911 he flew again over the mud flats at Newtownards, before crash landing. However, he flew again in October 1911 for the last time. At that stage he gave up flying probably having recognised the prohibitive costs of developing a commercially exploitable aircraft. However, his flights inspired James Martin, later Sir James, in his aeronautical interests, which are covered elsewhere in this volume.

These diversions strained Ferguson's relationship with his brother Joe and in 1911 he set up his own garage – May Street Motors which in 1912 became Harry Ferguson Ltd. His meticulous attention to detail and organisation quickly led to his business gaining an excellent reputation and acquiring several agencies for major car manufacturers. One of his early employees was William or Willie Sands, a natural born engineer and designer, who was to remain with him on and off, till the nineteen-fifties. While Ferguson was an ideas man Willie Sands was the man who converted the ideas into reality; one without the other would have been useless. Perhaps in retrospect Willie Sands did not get all the recognition he merited.

Another cornerstone of Ferguson's success was his wife Maureen (née Watson) from Newry whom he married in 1913. Like his mother she was quiet and serene, and she provided the support that he so much needed throughout his life.

Ferguson's company continued to prosper and on the outbreak of the First World War he took on the agency for the 'Overtime' one of the few farm tractors which had been developed by that time. He and Sands gave extensive public demonstrations to the ultra conservative Ulster farmers who were reluctant to accept mechanisation or the claimed economic advantage. As a result of the German U-boat attacks on UK shipping serious food shortages threatened the UK. Much of the farming manpower had been called up or had enlisted, and even many horses had been requisitioned by the army, so the only way to increase home production of food was mechanisation. In 1916 Ferguson was asked to promote the use of tractors during the 1917 ploughing season to increase food production in Ireland which was a major supplier to the rest of the UK. This brought him face to face with the practical difficulties involved in mechanisation besides the difficulties of overcoming the prejudices of the farming community.

His extensive war time experience of introducing tractors into Irish farming made him realise that they were simply being used as a mechanical horse; their potential value was not being realised.

Ferguson also recognised the danger posed by the stored energy in the rotating parts of the engine. When the plough struck a hidden obstruction, such as a rock or buried tree trunk, the forces generated were sufficient to cause the tractor to rear up and tip over backwards, sometimes injuring even killing the driver. He recognised that the solution was to design the plough as an integral part of the tractor, not as a conventional completely separate implement.

In 1917 Ferguson, with Sands, designed a two furrow plough for the Eros tractor which was based on a conversion kit for the Ford Model T car. But soon after Fords set up a tractor plant in Cork to produce their purpose designed Fordson tractor, which completely supplanted the Eros tractor. However Ferguson received some encouragement from Fords to adapt his plough concept to the

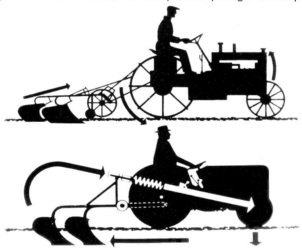

The principle of the Ferguson System.

Fordson tractor. In the early1920's Ferguson and Sands went to Ford's headquarters at Dearburn where a prototype of their plough for the Fordson tractor was successfully demonstrated to Henry Ford. Henry Ford was so impressed that he offered Ferguson employment which he declined. After further discussions they were unable to reach a mutually acceptable agreement for the marketing of the plough. Following further improvements in the design of the plough for the Fordson, Ferguson set up a company in America in 1925, Ferguson-Sherman Incorporated, to manufacture and market the plough. In 1928 Henry Ford ceased the production of the Fordson tractor in North America, and the Ferguson Sherman plough business was wound up.

In 1925 Ferguson having set up his American operation turned his mind to a much more radical approach in which not only could various implements be readily attached to the tractor, but it also incorporated automatic depth of work or draft control. The basic idea was an early form of automatic control or feed-back system. If the draft was too large then the implement was automatically raised to correct the error and vice versa. In practice Ferguson realised that the force in the lower supporting linkage was proportional to the draft. If this force was too great then this sent a signal to raise the linkage and vice versa. This was achieved by a hydraulic system in which the force activated a hydraulic valve to admit oil to a hydraulic cylinder to raise the implement or opened the hydraulic cylinder to the drain to lower the implement. An over-ride was provided so that the implement could be fully raised.

The patent for the draft control was granted in June 1926 and it was also filed in America. But having a patent does not mean that all the technical problems have been resolved. There were problems in developing what was, at the time, a very original hydraulic control system. It involved fitting an engine driven hydraulic pump and developing a hydraulic control valve activated by the force in the lower linkage. They experienced instability problems, which were not well understood at the time and they had oil aeration problems caused by over heating of the oil. Ferguson built a demonstration system on a Fordson tractor and, though several companies showed interest, he could not clinch a deal.

In 1932 Ferguson decided to build a prototype light weight tractor incorporating all his ideas including draft control. His concepts lent itself to a light weight tractor which had the advantage of minimizing soil damage by compaction. The tractor completed in 1933 was painted black and became known as the "Black Tractor". There were still teething troubles with the hydraulics but these were slowly overcome by further development.

The Black Tractor was demonstrated in England and Craven Wagon and Carriage Works showed initial interest but ultimately Ferguson clinched a deal with David Brown of David Brown Engineering, Huddersfield. A sales company, Harry Ferguson Ltd, was set up and the name of his Belfast business was changed to Harry Ferguson (Motors) Ltd, while David Brown set up a production company, David Brown Tractors. Production started in 1936 and

The Black Tractor

there was a good reception for the public demonstrations organised by Ferguson. However, the time was not ripe for its wide-scale acceptance by the farming community. This led to strains in the company and Brown initiated a market survey which suggested that the demand was for a larger tractor. Ferguson, however, insisted that the needs of small farmers world wide were for an even smaller and lighter tractor. This led to the parting of the ways.

In October 1938 Ferguson went to America and arranged to demonstrate the Ferguson-Brown tractor in a field close to Henry Ford's private residence. Clearly Henry Ford was immensely impressed and at a table set up in the Trial field he sat down with Ferguson. Ultimately they came to the famous gentlemen's agreement which was completed with a handshake without a formal legal agreement. The agreement was that Ford would manufacture the tractor, and Ferguson would be responsible for design and engineering as well as distribution.

On his return to England Ferguson found good reasons to withdraw from his partnership with David Brown on terms agreeable to both. In early 1939 accompanied by his family and design team he went to America. By March a new prototype tractor had been produced. Meanwhile the Ferguson-Sherman Manufacturing Company was reactivated to manufacture the implements together with a distribution and sales network. In 1941 this became Harry Ferguson Incorporated.

The famous handshake agreement.

Production of the tractor commenced in April 1939 followed by a public launch which received wide publicity and public acclaim. Despite a demonstration of the tractor in the UK in late 1939 Ferguson was unable to reach an agreement with the British Ford Motor Company to produce the tractor at their plant at Dagenham. However the Ford Ferguson tractor reached Britain under 'lease lend' agreement with America, and it became highly regarded.

Back in America despite problems and some inevitable friction, 90,000 tractors had been produced by 1 December 1941. As the war progressed Ferguson began to plan ahead and he launched his "World Plan" to alleviate poverty and malnutrition by the production of cheap food based on the world-wide exploitation of the Ferguson System. After the end of the war he returned to Britain where he convinced Sir Stafford Cripps, Chancellor of the Exchequer, to support the creation of a tractor plant in Great Britain. After tough negotiations he reached agreement with Sir John Black of the Standard Motor Company to set up a tractor plant in an ex-shadow factory in Banner Lane in Coventry. Production went reasonably smoothly and late in 1946 the first of many hundreds of thousands of Ferguson System tractors, the well loved "Fergie", rolled off the production line. In 1953 Ferguson came to a gentlemen's agreement to sell Harry Ferguson Incorporated, the Detroit Company, and Harry Ferguson Ltd, based in Coventry, to Massey Harris, which became Massey-Harris-Ferguson of which he became the Chairman for a short time. In 1954 he resigned and sold his shares in Massey-Harris-Ferguson.

Harry Ferguson driving a 'Fergie; down the steps of Claridge's Hotel 1948.

Henry Ford retired in 1943 and his relatively inexperienced grandson Henry Ford II became President. At the end of the war Fords suffered heavy losses, not least of all in the production of tractors. In 1946, Henry Ford II, after unsuccessfully trying to negotiate a written agreement with Ferguson, established a new distribution company the Dearborn Motor Company. This replaced Harry Ferguson Incorporated. When Fords introduced a new tractor which incorporated an unmodified Ferguson System of hydraulics and linkage, it was the last straw and Ferguson initiated legal proceedings claiming a modest $251 million in damages. These proceedings dragged on until April 1952 when he settled for $9.25 million; so much for a gentlemen's agreement. But perhaps neither Henry Ford nor Harry Ferguson were gentlemen!

In 1950 in his declining years Ferguson set up Harry Ferguson Research to develop a revolutionary four wheel drive vehicle. This gave rise to wide publicity and interest by car manufacturers, but it never reached fruition. Increasingly he suffered bouts of severe depression, which had dogged him for most of his life, and other medical problems. On 25 October 1960 Ferguson was found drowned in his bath and a post mortem examination found that he had taken an overdose of drugs. At the inquest the jury returned an open verdict. He predeceased his wife Maureen who lived on for another five years. They had one daughter, Betty, born in 1918.

When the history of the twentieth century is written the Harry Ferguson's tractor system will be recognised as one of the greatest inventions amongst a galaxy of inventions such as aeroplanes, radio, television, computers and space exploration, to mention but a few. It has enabled agriculture to greatly increase food production and reduce the cost of food, without which many more of the ever increasing world population would be faced with starvation.

Sources:

1. 'Invention, Innovation and Design' by G.B.R. Feilden, First Harry Ferguson Memorial Lecture, pub The Queen's University of Belfast, 1970.

2. 'Harry Ferguson – Inventor and Pioneer' by Colin Fraser, pub. John Murray, 1970. ISBN 0 7195 2660 4.

3. "Harry Ferguson" by Bill Martin, pub. Ulster Folk and Transport Museum, 1984.

SIR EDWARD J. HARLAND

Edward Harland was born at Scarborough in May 1831 the sixth child of a family of eight. His father was Dr William Harland born in Rosedale who studied medicine at Edinburgh and practiced in Scarborough. Dr Harland had a love of mechanical pursuits and mixed with some of the great engineers of his day and age. In 1827 he invented and patented a steam-carriage embracing a multi-tube compact boiler and plate condenser and he produced a working model which worked satisfactorily. Edward's mother was the daughter of Gawan Pierson of Goatland near Rosedale. For her day she was surprisingly interested in mechanical things, and assisted her husband in preparing many of his plans.

As a youngster Edward was brought up to be interested in practical things and model making and in the workshops and boat building yards in Scarborough. He attended the local Grammar School, where he was particularly interested in drawing, geometry and Euclid. At the age of twelve he went to the Edinburgh Academy, where he was near his eldest brother who was studying medicine. Shortly after his mother's death in 1844 he returned to Scarborough where he received tuition from an old school master. Though his father wanted him to be a barrister he was determined to become an engineer. So his father, who knew the famous locomotive engineer George Stephenson, arranged for him to be apprenticed to the great engineering works of Robert Stephenson and Co of Newcastle-upon-Tyne. He started his five year apprenticeship on his fifteenth birthday.

After four years of workshop floor experience Harland joined the drawing office where he obtained design experience. On the completion of his apprenticeship at the age of twenty there was a considerable recession, which unsettled him, so he resigned and spent a couple of months in London. It was the time of the 1851 Great Exhibition and he spent much of his time feasting his eyes on the enormous range of innovative machines exhibited, as well as the art of the Victorian age.

The sister of Edward Harland's aunt was married to Gustav Christian Schwabe who was involved with a shipping company in Liverpool and in the 1840's became a junior partner in John Bibby and Sons a Liverpool shipping company. Gustav Schwabe provided a letter of introduction to Messrs J and G Thomson, marine engineer-builders in Glasgow, where Edward gained further experience. During his stay in Glasgow he assisted Messrs Bibby in the purchase of a steamer. When Messrs Thomson decided to build their own vessels he assisted the naval draughtsman. When the former left, Harland was promoted to fill his place.

In 1853, after little more than two years with Messrs Thomson, Edward Harland accepted the post of manager with Mr Thomas Toward, a shipbuilder on the Tyne, where he supervised the building of ships and boilers. He recognised the great importance of quality control in achieving commercial success and he would not tolerate poor workmanship. The declining health of Mr Toward led to Harland being in full charge but he felt that the prospects were not good. In late 1854 he took up the post of manager of the shipbuilding yard of Robert Hickson and Co in Belfast. It was a recently developed yard on the Queen's Island, an island created during the straightening of the channel of the River Lagan by William Dargan. This was the beginning of his life long association with Belfast and in particular Queen's Island.

In this new post he had complete responsibility for the building of ships and he quickly discovered that the quality and quantity of work output was unacceptable. Following Toward's death (1855) he recruited the Head Foreman and several leading hands who came to Belfast and developed the shipyard to his own systems and satisfaction. After the successful completion of orders in hand to the satisfaction of the owners, new orders were forthcoming for several large sailing ships and screw propelled ships. In 1857 he recruited Gustav Wilhelm Wolff as his assistant. Wolff, a German, was the son of Gustav Schabe's sister.

After three years at Hickson's yard Edward decided that he wanted to start his own yard and when Mr Hickson learned of this he offered to sell him his yard. With the assistance of Gustav Schwabe, the sale was completed in 1858, and the yard was renamed Edward James Harland and Company. This was immediately followed by an order for three screw propelled steamers for J. Bibby Sons and Co. with engines from MacNab and Co of Greenock. These were the first of a succession of orders from J. Bibby, Sons and Co, which firmly established the shipyard as a major player in the world of ship building.

At this juncture Edward Harland found time in his very busy life to court and marry Rosa M. Wann of Belfast on the 26 January 1860. She was to prove to be a most supportive wife with sound judgment and business sense. They had no children.

The successful completion of the steamers for J. Bibby, Sons and Co., brought a further order for six larger vessels. In 1861 this increased work load led Edward to take on Gustav Wilhelm Wolff as a partner to form Harland and Wolff (H&W) which was to develop into the "shipbuilders to the world".

When Bibby ordered their first steamers they were built to the specification they laid down. But with the next two vessels of 1,854 gross tons Harland adopted a radically different design. He proposed to construct these vessels with greater length to accommodate the increased tonnage, while not increasing the beam or breadth and with a flat bottom. This required very little increase in power to achieve the same speed. To achieve the necessary strength he made the upper deck entirely of iron thus creating what was essentially a very strong box girder. He also installed Hall's design of surface condenser, which greatly increased the thermal efficiency, giving a 20% improvement in fuel consumption. Despite the sceptics, who referred to these vessels as "Bibby Coffins", they proved to be highly successful; by 1870 H&W had built eighteen ships for J. Bibby, Sons and Co. including ships up to 3,113 gross tons.

In 1868 Gustav Schwabe helped Thomas Henry Ismay of Liverpool to acquire the White Star Line. Subsequently in 1869 Gustav Schwabe was responsible for encouraging Ismay to set up the Oceanic Steam Navigation Company to compete on the rapidly developing Atlantic trade routes. He also organised the necessary financial support on the understanding that Harland and Wolff built the ships needed. As a result six transatlantic steam ships of 3,800 gross tons were ordered. This required a major modernisation of the yard and set a pattern for the next forty years of regular refurbishment and improvement.

Considerable thought was given to saving fuel. A member of staff made voyages on steam ships fitted with the newly developed compound engine. As a result it was decided to adopt this design, and engines were ordered from Maudsley, Sons and Field of London and George Forrester and Co of Liverpool. The ships were designed to provide unheard of comfort with gas lighting, commodious smoking saloon, electric bells, a promenade deck, etc. These were to become the first recognisable transatlantic liners which for the first time made the travel experience enjoyable and even delightful.

The first of the six ships, the *Oceanic*, started the transatlantic traffic for the company flying the White Star Line flag on March 1871. Though it experienced severe weather the ship made a splendid voyage with a heavy cargo of goods and passengers. This was the beginning of a long association with the Oceanic Steam Navigation Company culminating with the construction of three sister ships the *Olympic*, *Titanic* and *Britannic* before the First World War.

SS *Oceanic*

During its existence from 1853 the yard had grown from employing 48 men to over 1000. In 1874 Edward Harland had begun to broaden his interests including becoming Chairman of the Belfast

The Engine Works

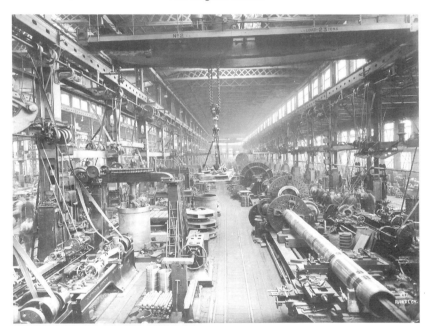

Harbour Commissioners, while Gustav Wolff, who had joined the newly created Belfast Ropeworks Company as Director, became its Chairman. So it was timely for the partnership to be broadened by embracing William J. Pirrie, Walter H. Wilson and Alexander B. Wilson, who resigned in 1877. From that time Edward Harland began to reduce his commitments to the partnership, while increasing his outside interests.

Up till the end of the end of the eighteen seventies, H&W had relied on the supply of engines from the independent marine engine-builders. In 1879 the decision was reached to invest a massive sum in an engine and boiler works in Queen's Road, which produced the first set of engines in 1881. The engine works were to become one of the world's most major marine engine-builders. At the same time the yard began to use Bessemer steel in place of wrought iron in the construction of vessels which resulted in significant saving of weight. Edward Harland had experimented with open-hearth steel as early as 1869 but had found it to be too brittle. There had also been many other developments, such as the reduction in the corrosion of propellers by using zinc plates attached to the hull in the vicinity of the propeller.

In 1884, the year in which H&W made their first triple expansion steam engines, Edward Harland at the age of fifty three withdrew from active involvement in the company though he continued making loans to it. During the rest of his life he pursued his other business interests, and became much involved in local and national politics. In 1885 and 1886 he was Mayor of Belfast and in 1887 High Sheriff of Co Down. Following the very successful visit of the Prince and Princess of Wales, while he was Mayor, he received his Baronetcy in 1885. From 1889 through to his death in 1895 he was the Unionist MP for North Belfast and very opposed to Home Rule for Ireland which he feared would greatly harm the business prospects for H&W and other companies in Belfast.

Sir Edward Harland died in his sixty fifth year on the 24 December 1895 and he was buried on the 28 December. On the day of his funeral all the flags on buildings and ships in Belfast were at half mast, and his funeral cortege was attended by huge crowds and it was headed by five hundred employees of H&W marching four abreast. Everybody recognised his enormous contribution to the development and prosperity of Belfast by the creation of a major world shipbuilding yard and engine works.

Sources:

1. 'Shipbuilding in Belfast - its Origin and Progress' by E.J. Harland, Engineer and Shipbuilder, in "Men of Invention and Industry" by Samuel Smiles, pub. John Murray, London, 1884.

2. 'Shipbuilders to the World - 125 Years of Harland and Wolff, Belfast 1861-1986', by Michael Moss and John R. Hume, pub. The Blackstaff Press, 1986, ISBN 0 85640 343 1.

SIR CHARLES LANYON

Charles Lanyon was born in 1813, the youngest of three sons of John Jenkinson Lanyon and Catherine Anne Mortimer of Eastbourne, Sussex. John Lanyon, whose family was originally of Cornish stock, was a purser in the Royal Navy and his wife was from an old family of minor landed gentry who had been prominent in Sussex since the Wars of the Roses. Their influence was profound. From his father he learned the skills of management and leadership, punctuality, the importance of attention to detail and the benefit of hard work. From his mother he acquired the social skills and pleasant manners which, along with his good looks, gave him an entrée into the society of the day.

His work as an architect was probably influenced by the familiar surroundings of his childhood. His father was churchwarden of St Mary the Virgin in Eastbourne, which was a mediaeval building of true gothic design. The village was then set back from the sea in a fold of the South Downs and did not become the prosperous sea-side resort and epitome of classical elegance until the middle of the century. The Mortimer and Lanyon families had friends and relations along the coast, in Brighton, where the young Charles was probably more impressed by the regular stucco facades and neo-classical doorways of the sweeping terraces than the Hindu-Gothic extravagance of the Pavilion. He would also have been exposed to the contemporary work of the rising Charles Barry who later designed the new Houses of Parliament in London.

Because the Lanyon family was not sufficiently wealthy for the members to be considered landed gentry, each of the sons was required to find himself an occupation. One brother emigrated to Australia where he built up a successful sheep ranch the other joined the Royal Navy. Charles was trained as a civil engineer being articled to Jacob Owen, an experienced clerk of works in the Royal Engineers Department of Ordnance in Portsmouth. During this apprenticeship, Owen was appointed Engineer and Architect to the Board of Works in Dublin. The young Charles Lanyon went with him and completed his apprenticeship in the Irish capital.

In 1835, there was a reorganisation of local government and the post of County Surveyor was established. An examination or competition was held for these new posts and Charles Lanyon took second place. As a result, he was appointed to County Kildare. This put him in a position to marry Helen Elizabeth Owen, his master's daughter.

Within a year of this reorganisation, Thomas Woodhouse, who had taken the post of County Surveyor for County Antrim, decided to concentrate on building railways and took up the post of resident engineer to the Midland Counties Railway in England. Lanyon applied for the vacant position of County Surveyor. He was successful and therefore, in 1836, he moved to Belfast where he spent the remainder of his life helping transform the bustling market town into a thriving industrial city.

Lanyon's impact on County Antrim was considerable as he convinced the Grand Jury, by whom he was employed, to appoint twelve assistant or Barony surveyors. With this team he rapidly established a reputation for efficiency. He undertook a wide variety of tasks with speed and reliability and raised the tempo of the local architectural and civil engineering professional practice.

The Grand Jury represented the Establishment and the landowners, many of whom were also involved in the growing industrial wealth of the town and county. Lanyon's background and social standing combined with his good manners and exquisite taste appealed to the country gentlemen. His businesslike reports and his practical, problem solving, approach to professional matters appealed to the newly rich businessmen. It was soon apparent that while the county's affairs were running smoothly, which Lanyon was capable of ensuring, there was no objection to the County Surveyor undertaking private commissions. The result was that by the time he reached the age of thirty, he had become the most prominent architect in the province, not just the county.

He ensured that his works were constantly in the public eye as a means of gaining these outside commissions. He would undertake charitable works free of charge in the knowledge that the patrons and managers would then use his services on a regular basis for more expensive works. Whether this was a matter of luck or calculation is open to debate but human nature is such that the latter is the more likely proposition. The outcome was that in the first fifteen years after his arrival in Belfast he was responsible for roads and bridges, public and private housing, churches and schools, banks, charitable institutions, railway tracks and stations, water supplies and sewerage. In this period, there was a distinct change in the local architectural patronage in Ulster. Before he arrived most major private building projects were designed by architects from outside the province, mainly from Dublin, Edinburgh or London. By 1850, it was quite unusual to find an outside architect employed for such tasks.

The two major road improvements which Lanyon made in his first years as County Surveyor in Antrim were the Frocess Road and the Antrim Coast Road. They required very two different construction techniques. The former crossed a part of the Frocess Bog, between Ballymena and Ballymoney, over a very unstable surface and required the planting of two lines of Douglas Fir trees the roots of which intermingle and provide a base for the road and which survive as a form of memorial. The Coast Road, which has been described as "second only to the Grand Cornice above Monte Carlo", required the removal of large quantities of rock and the building of sea walls, work which was carried out by subcontractors to Lanyon's designs. The Glendun Road Viaduct, a part of the road which is unseen by travellers, shows the confidence and ability of the young Lanyon.

Through his designs it appears that Lanyon had firm views in the matter of style. He was more inclined to follow the Victorian reaction to the Regency frills and frivolity. Therefore, each of his buildings or bridges were more likely to show the adaptation of an historical style to its modern purpose. Churches adapted medieval and gothic forms because this is what he saw as the Christian style of worship. For the same reason his private houses took their form from the status rather than the taste of the client and echoed themes of baronial castles, mediaeval manors or merchant palaces. In Institutional and Educational buildings, the Tudor - Jacobean style which marked the great age of the foundation of almshouses and colleges can be readily identified. Courthouses were classical and prisons foreboding. Road bridges and railway viaducts can have their design influences traced back to the great roman aqueducts of even greater antiquity.

He was presented with a silver salver by the Church Accommodation Society to mark his work, having designed fourteen places of worship between 1838 and 1843. The designs for these churches were given freely to the Society. They are mainly in the Gothic Revival style, examples being St John's Kilwarlin, Hillsborough (1840) St John's Glynn, (1841) St John's Whitehouse (1842) and Glenoe (1842). Having been built in a very Presbyterian area, this latter had the distinction of stopping the fighting between Trinitarian and Unitarians as they joined common cause against the Episcopalians. He also designed Trinity Church, Kirkcubbin (1843) as one of very few in a Greek Revival style. Another in this style is the First Non -Subscribing Presbyterian Church in Holywood which dates from 1849.

The Deaf and Dumb Institute, Belfast, which gives design clues to the later Queen's College.

Of the many buildings designed in the Tudor Revival style, The Queen's College building of 1849 is the most famous. The central tower is a free translation of the Founder's Tower of Magdalen College, Oxford. This theme is repeated in the smaller towers which break the long and strictly symmetrical front. The ambulatory and other features in the rear aspect of the building are not as highly decorated as the front because of the problems with funding experienced even then by the college authorities. Before this, the Ulster Institute for the Deaf and the Dumb and the Blind on the Lisburn Road, which was opened in 1845, was said to be a rehearsal for the major work. Both were build of red bricks but the earlier one suffered more seriously from deterioration and was pulled down in 1965.

Several major houses were also built in the Tudor or Jacobean renaissance style. Stranmillis House was built in 1858 for Thomas A. Batt, a director of the Belfast Bank. It was home to other important industrialists including Sir Thomas Dixon and Walter H. Wilson of Harland and Wolff, before being converted into the Teacher Training College. Kintullagh, Ballymena was built in 1862 for William A. Young and is now St Louis Convent. Gills Almshouses, Governors Place, Carrickfergus also suffered deterioration of their bricks and have been rendered in an effort to conserve them.

Lanyon's own painting of the original Queen's College concept.

As County surveyor, Lanyon was responsible for public buildings and bridges in Belfast as it was not then of city status. Thus, the Crumlin Road Goal and the County Courthouse directly opposite were his responsibility. When the jail was built in 1846, it was linked by an underground passage to the Courthouse although the latter did not come into use until 1850. Commentators have noted that the furnishing of the court seemed to draw on Lanyon's experience with churches and lecture halls. His other impressive work, the Queens Bridge was opened in 1843 but had to be widened with cantilevered footpaths, cast iron railings and gas lamps in 1885.

Lanyon also made use of the Italianate style in many buildings. Assembly's College of 1855 and the Customs House of 1857 are two important examples. He had used this style when reworking Ballywalter House for Andrew Mulholland in 1847. The Abbey, Whiteabbey and Craigdarragh, Helen's Bay are other examples of private houses in this style. At the same time, he was also designing in the Gothic revival style illustrated by the courthouse in Ballymena (1844) with the adjacent Adair Arms Hotel and the covered walkway between to give the judge access between bed and bench.

In1860 he resigned from the position of County Surveyor to concentrate on his private practice and other interests. In 1862 he was both Mayor of Belfast and President of the Royal Society of Irish Architects. He represented Belfast in Parliament from 1866 to 1868 and was knighted in that year. While a member, he served on the Select Committee on Scientific Instruction. This committee had significant influence as its deliberation led to the Education Act of 1871 which introduced universal education. Due to divisions within the Conservative Party, Lanyon lost his seat at Westminster. Lanyon represented St George's ward for many years (1861 - 1871) on the Belfast Town Council and was a Harbour commissioner from 1862 to 1886. He was also a Deputy Lieutenant for County Antrim and High Sheriff in 1876. Among his many other interests, he was a Trustee of the Dioceses of Down and Dromore, president of the Board of Management of the Belfast Government School of Art, an honorary secretary of the Ulster Institute for the Deaf and the Dumb and a member of the Board of Management of the Belfast Nurses' Home and Training School. As a result of his work for the local railway companies, Lanyon was a director of several. His various posts included chairman of the Belfast, Holywood and Bangor Railway Company, vice-chairman of the Belfast and Northern Counties Railway Company and a director of the Carrickfergus and Larne Railway Company, as well as being a director of the Blackstaff Flax Spinning Company. He was a committed Freemason rising to the distinction of Most Worshipful Grand Master of the Freemasons of Ireland.

In 1860, Sir Charles Lanyon took as partner in his practice William Henry Lynn who had joined him as an apprentice in 1845. His son, John Lanyon was made a partner in 1864 and the firm was known as Lanyon, Lynn and Lanyon until 1872 when Lynn set up on his own account. Another talented apprentice, Thomas Turner, moved to Londonderry where he built his own practice. After about 1855 it is thought that many of the buildings designed in Lanyon's offices were the work of these talented staff. Some of his buildings have been pulled down and some, such as Shane's Castle, which he redesigned in 1862, were destroyed by terrorist activity while the Randalstown Viaduct (1856) has been disused since the railway line was closed in 1955.

Amid all his activities, he was an active church member. In his later years he lived at The Abbey, Whiteabbey, a house he had originally designed for Richard Davidson MP, and worshipped at Carnmoney, where he was a churchwarden. When he died, on 31 May 1889, he was buried in Newtownbreda Churchyard in a mausoleum which is now a listed building. His wife had predeceased him as had two of their five sons and one of their four daughters.

Sources:

1.	Belfast Telegraph

JAMES MACKIE, Jnr

James Mackie Jnr. was born in College Gardens, Belfast, in 1864. He was the second child of a family of three - an elder sister, Elizabeth (Lizzie) and a younger brother, Thomas (Tom). Their father, James Mackie, was born in Dumfries, Ayrshire in 1823, and he married Mary Miller of Broughshane near Ballymena in 1860.

James Mackie Snr served an apprenticeship with a steam engine builder in Glasgow. In 1843 he came to Ireland to erect a steam engine in St Mary's Flax Mill in Drogheda. Subsequently, he became the mill engineer at Richardson's Linen Mill in Bessbrook, before joining Scrimegour in the Albert Street Foundary in Belfast as works manager in 1845. When Scrimegour became bankrupt in the following year, James bought the assets and carried on the business. This was the beginning of what was to develop into the world renowned company of James Mackie and Sons Ltd.

James Mackie Jnr turned his back on what might have been a promising academic career and at the age of fourteen he joined his father's company. He gained experience going round the linen and textile mills in Northern Ireland repairing and modifying their machines, as well as making contact with both the owners and the operatives. This led to his strong conviction of the importance of his family having a shop floor apprenticeship as the best passport to success. This became an accepted family tradition up to the demise of the company. By the age of twenty James was recognised by his clients as an ingenious and dependable businessman.

When his father died in 1887 James, then aged twenty-three, took over the management of the company with the support of his mother, sister and brother. The partnership, known as James Mackie and Sons, started to concentrate on the design and building of spinning and twisting frames for the flourishing linen industry. While 'The Boss', as James became known, drummed up business at home and abroad, his mother ran the office and his brother looked after the factory.

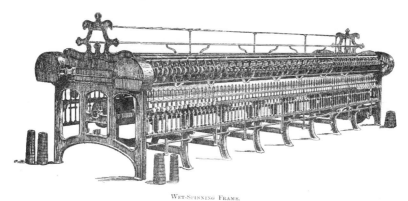

Wet Spinning Frame

Despite the competition with other well established companies in Northern Ireland and the rest of the UK, the business prospered. This necessitated the removal of the business in 1893 to a greenfield site on the Springfield Road in West Belfast. The Belfast Banking Company provided a substantial loan for this development and over the years it continued to give the company support when required. The Boss recognised the cyclic nature of the textile machinery trade, so a flexible design was adopted for the new building to enable it to be used as a flax spinning mill if there was a collapse in the machinery trade.

In 1894 "the Boss" married Elizabeth, 'Lily', Pringle, daughter of Alexander Pringle the manager of the Bessbrook Mill. They had six children - Jim, Jack, Fraser, Isobel, Grenville and Lavens, who were all to play a part in the family business except for Isobel. Around 1904 they moved to Hazelbank on the shore of Belfast Lough at

Whiteabbey. Lily, besides being a devoted mother, played a crucial supporting role for James in his business by entertaining the many clients who visited the company from home and abroad.

In 1909 during a turn down in the textile machinery trade the Boss approached Sir Otto Jaffe who owned a brickworks, and encouraged him to build a large spinning mill, for which Mackies provided the machinery. This became the Strand Spinning Company in 1912, which in its day was the largest flax tow spinning mill in the world. In 1920 the Boss sent his son Fraser to run the Strand.

The Boss travelled widely to drum up business at a time when travel demanded much more stamina than at the present time. His diary records some of these journeys such as one to Russia in 1899 to visit the Milehess Mill to the East of the Volga. His journey in late winter involved a long train ride from Moscow, followed by a horse driven sledge journey with an overnight stay in rudimentary accommodation. Clearly one needed to be pretty tough to maintain contact with existing customers and to get new customers.

These business trips, as with other members of the family, not only drummed up orders for existing designs but generated a demand for completely new machines. The Boss having had extensive practical experience was able to transmit ideas for modification of existing designs and for new designs, to the design office on his return to Belfast. This gave the edge over their competitors.

He was again in Russia on the outbreak of the First World War. The war led to the loss of supply of French and Belgian flax used in the mills in Northern Ireland. His clients back home commissioned him to purchase several thousand tons of high quality flax from around Smolensk. The question was how to transport it back home with the Baltic closed to shipping? The story goes that he went by rail to Odessa on the Black Sea, but there were no Russian steamers available. But there were two old Turkish warships tied up in the harbour, which he purchased. No sooner had they set sail through the Dardanelles than Turkey joined the War on the side of Germany. When the British Navy saw two Turkish warships steaming out into the Mediterranean they promptly sank them, and James lost a fortune.

By the time the first war broke out in 1914 James Mackie and Sons had expanded to employ a workforce of six hundred. During the war the factory was turned over to the production of shells used in massive quantities in the battles on the Western Front. For the first time in engineering manufacture in the British Isles, women were employed on both the day and night shifts as machine operatives. This is immortalized by the well known and admired Belfast artist, William Connor in his painting of a Mackie's employee, Ellie Grey, operating a lathe wearing a long skirt and apron.

After the war James along with other leaders in the linen industry recognised the need for collaborative research to maintain the competitiveness of linen. As a consequence the Linen Industry Research Association (LIRA) was set up in 1919 at Lambeg near Lisburn. It was one of the first of many industrial research associations set up in the interwar years in the UK and the only one based in Northern Ireland. This led to many important developments over the next fifty years not only in linen but also other natural and man-made fibres.

In 1924, with encouragement from the Ludlow Company of Boston, the first experiments were carried out on the design of jute spinning machinery based on Schneider's patents. This ultimately led to James Mackie and Sons dominating the world market for jute spinning machinery. It also led to Connswater Spinning being set up by Mackies in 1949 when jute spinning machinery for a new sack mill for the abortive British Government's groundnut project in Tanzania was no longer needed. In 1954 the Lagan Jute Machinery Company was set up in Calcutta, which in 1976 was sold to the Indian Government. That was not to be the end of the saga as in 2000 the Lagan Jute Machinery Company was sold to the Murlidhar Ratanlal Exports Ltd of Calcutta in which Gordon Mackie, the son of Lavens M. Mackie is a Director.

James Mackie Jnr received many honours in his lifetime. He became deputy Lord Lieutenant of County Antrim, a J.P., Chairman of the Belfast Chamber of Commerce and a Belfast Harbour Commissioner. After 1933 progressive arterial disease gradually wiped out his memory and most of his mental faculties, and he died in 1943. His wife, Lily, predeceased him in 1940, while his younger brother, Thomas F. Mackie, lived on till 1956.

Under the Boss the company, despite intense local and national competition, had prospered and developed into a major internationally recognised company. His five sons had joined the business during the First World War. When he was forced by ill health to retire his son, Jim, became Chairman, with Jack and Lavens as Joint Managing Directors and his nephew Stuart Mackie in charge of the main works, while Fraser remained in the control of the Strand Spinning Company.

During the Second World War the company was again mainly engaged in a wide range of armaments production, including armour piercing shot, gun barrels, Stirling bomber fuselages and Sunderland flying boat tail fins. There were altogether fourteen factories employing 12000 people of which sixty percent were women. Their contribution to the war effort was enormous. At the end of the war Jim was offered a knighthood for wartime services and when he replied that he would be delighted on condition that his four brothers were also knighted, the establishment was outraged. Their father had drummed into them that one of the main reasons for the success of the company depended on his belief that it was one for all or all for one.

After the second world war there followed fifteen immensely successful and profitable years with the design and production of new and more productive textile machines. But from 1960 there was a steady decline in the business for many reasons. These included introduction of cheaper synthetic fibres, high taxation on private companies, the rapid decline in the textile industry in the Western World, and from 1968 political unrest, particularly in West Belfast where the company was situated.

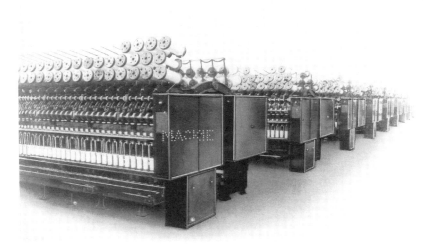

Dry Spinning Frame

In 1976 the Mackie family withdrew from ownership of the company having planned a tax efficient strategy and the factory complete with its lands, equipment, patents, etc, was handed over to a trust on behalf of the workforce. With little working capital, the continuing decline of the textile industry, the lack of Government grants available to other companies and the continuing political unrest, the business continued to decline. In 1989 Leslie and Gordon Mackie who were the last of the family still employed in the company left. After various vicissitudes, including two changes of ownership, the company overburdened by debt was declared bankrupt and closed in 1998.

James Mackie Jnr was largely instrumental in the development of James Mackie and Sons into a successful international textile machinery company. The family cohesion he engendered, with his emphasis on the importance of practical workshop training for the members of the family who joined the company, was important in his day and age. Perhaps now, with the high rate of technological change in recent decades, leaders of such companies also need a high level of technical education integrated with practical training. In the last analysis, the decline of so many of our great engineering companies in the UK reflects the belief of our political masters that the way ahead is the service industries and information technology. Time will show if this belief was well founded or disastrous.

SIR JAMES MARTIN, CBE

James Martin was born on the family farm near Crossgar, Co. Down on the 11 September 1893. His father, Thomas, came from a long line of farmers in Co. Down. He had an interest in machines and had his own workshop, where he maintained and constructed machines for his own farm and also those in his locality.

When James was two his father died and he was brought up in a household dominated by women - his mother Sarah, her mother-in-law and sister-in-law with James's sister, Jane two years older. At the local school which he attended up to the age of fourteen he excelled in how things worked and at home he was an avid reader of technical books. As the only man in the household he was expected to take over the farm, but he showed little interest in farming. Instead he was much more interested in his father's workshop. His mother accepted his interest in engineering and saw his future as a civil engineer. She arranged for him to be interviewed by Professor Frederick Fitzgerald who was then Professor of Civil Engineering in the Queen's University of Belfast. He opined that James was a practically minded young man more suited to practical engineering than engineering science. No doubt James was relieved, and in later life it must have given him a wry satisfaction when he was awarded an Honorary DSc by the Queen's University.

James taught himself to use a lathe and other tools in his father's workshop and he also learned from observing and talking with skilled mechanics in the thriving industrial city of Belfast. At an early age he began designing and making things for himself. Though he could, no doubt, have got an apprenticeship one guesses that, besides being committed to helping out his mother on the farm, he wanted to be his own man and do his own thing.

From 1908, when he left school, until 1919 he remained at home helping his mother but increasingly devoting his time providing engineering advice and services to local farmers and others. Like many young men of his time he was interested in internal combustion engines, cars and aeroplanes, which were at the leading edge of technology at that time. He visited the premises in Belfast where Harry Ferguson was building his own aeroplane and he was present on the 31 December 1909 when Harry Ferguson made the first powered flight in Ireland. That is what probably stimulated James's life long interest in aviation. Subsequently, in 1911 James made his first invention of a speed indicator which warned a pilot when his airspeed was too high or low. He made a patent application, but it was never registered.

In 1911 James's sister left home to make her own way in life and in 1919 she married Edwin Burrell and settled in London. James's relationship with his sister was very close and in 1919 he left home and found lodgings and a workshop in Acton near to his sister. When he ran out of money he left his lodgings and moved in with his sister and brother-in-law for the next twenty years.

During the following years his business slowly developed as people began to recognise his innate abilities and inventiveness. He began to build one-off specialised vehicles and in 1923 he applied for a patent for a new design for body-frames of vehicles which was granted in 1925. This was a forerunner of the design he was to adopt in airframe construction. A further patent with John Campbell, registered in 1927, was related more specifically to improved methods of construction of vehicle body frames, aircraft fuselage frames and other conveyers for transporting goods. In 1929 James applied for another patent which extended his ideas to aeroplane wings and spars.

Even by the end of the ninety-twenties James was only employing three, but his income allowed him to buy a car which he modified for competitive racing. However this came to an ignominious end when he crashed into the butcher's shop in Comber in the 1928 Ulster Tourist Trophy race. In 1928 he started to take flying lessons at the London Aero Club at Stag Lane under a flying instructor, Valentine Baker.

It was at this time James decided that his real aspiration was to design and build aeroplanes but clearly his existing workshop was inadequate. In the meanwhile his sister had moved to Gerrards Cross in Buckinghamshire where James found a suitable disused factory with space for expansion. His brother-in-law purchased it and leased it to him. He moved into the factory in 1929, which was named the Martin's Aircraft Works, where he hoped to exploit his patents on airframe construction and a more recent joint patent with Percy Waterman Pitt related to propeller shafts.

Without any capital it was fortunate that Pitt had sufficient finances to back the construction of a twin seater aircraft which James envisaged would be flown by Miss Amy Johnson. She had also been a flying pupil of Valentine Baker. She was the first aviator to fly solo from London to Melbourne in 1930, a flight suggested her by James. Good progress was made on constructing James's first aircraft until, as a result of the 1930's depression, Pitt was forced to withdraw financial support.

About this time Valentine Baker acquired a new pupil, Francis Francis, who had inherited a fortune from his grandfather one of the founders of the Standard Oil Company. When Valentine introduced Francis to James they immediately took to one another. Francis was enthusiastic in support of James's aeroplane project, which was ultimately to be designated the Martin-Baker I. As progress was made Francis decided to formalise the relationship by encouraging James and Valentine to set up a company, which he would finance. The Martin-Baker Aircraft Co Ltd, incorporated on the 17 August 1934, took over the Martin's Aircraft Works.

The company was still small with only six employees. It was wildly optimistic to think they could design and build an aircraft, even in the nineteen-thirties when aircraft were still relatively simple. More especially as James was the sole designer and a born perfectionist making it impossible for him to firm-up on a final design. However he had the considerable advantage of a very loyal and dedicated workforce.

The MB1 was not completed until the spring of 1935 when it was test flown by Baker; it passed with flying colours. But in the aftermath of the depression the climate for a new venture was poor and, despite excellent publicity, it did not attract orders. This might have discouraged many another but not James who was a positive thinker. He decided to build a fighter aircraft with Francis's financial backing. The MB2 was a single seater eight gun fighter with a fixed undercarriage. In its design James adopted a unit construction, which simplified construction, and he gave a lot of thought to accessibility, maintainability and safety. When the Ministry of Aircraft Production would not release a Rolls Royce Merlin engine he adopted a Napier Dagger III.

The MB2 was flight tested by Baker on the 3 August 1938 but modifications were needed to improve its flying qualities. In November the Ministry carried out its own test programme. Their report was very favourable on the features such as accessibility, maintainability and the speed of arming its eight guns. However, they found its flying qualities were unsatisfactory as far as providing a stable platform for accurate fire of its guns. Nevertheless they purchased the machine for further testing.

It is probable that the Ministry saw Martin-Baker as being too small to be taken seriously as an aircraft manufacturer. But they did not want to discourage James as they recognised his innovative abilities as a valuable asset in time of war. For example James's design of blast tubes, to vent the gases from the guns in the MB2, were adopted for the Hurricane fighter and other aircraft. In 1938 James had developed an explosively activated cable cutter for fitting to the wings of aircraft to cut the cables of barrage balloons. These saved the lives of many aircrew and also many aircraft.

James, however, was still set on producing a single engine fighter and in 1939 he came up with the proposal for the MB3 with three 20 mm cannon in each wing. It was initially designed round a new Rolls Royce engine, the Griffon, currently under development. Ultimately, in March 1940, the Ministry decided that the aircraft should be designed around a new and unproven engine, the Napier Sabre. With numerous demands, such as a jettisonable canopy for the Spitfire and the twelve machine gun mounting for the Havoc night fighter, James was left with little time to devote to the MB3.

It was not until the 31 August 1942 that Baker test flew the MB3. Over the next two weeks its handling characteristic proved to be excellent but there were problems with the Sabre engine. Then, on the 15 September, the engine failed completely just after take off and the aircraft crashed killing Baker. It was a devastating loss to James of his dearest friend and partner. He blamed himself for not having stuck out for a Griffon engine.

Martin-Baker was still under contract to produce two prototypes and, though the Ministry probably did not expect the aircraft to go into production, they were still interested in its many novel features. In the end it was agreed to produce a second prototype to be fitted with a Griffon engine which incorporated so many other changes that it was designated the MB5. It first flew on the 12 May 1944 but substantial changes were called for. It next made a spectacular debut in October 1945 at the Farnborough 'At Home' three day event but an engine failure led to a forced landing. It was ultimately delivered for formal trials in February 1946 which it passed with flying colours in every respect. Many considered it the best machine developed during the war, even if produced on a shoestring. But with the end of the war, and the advent of the jet fighter, the days of the piston engined aircraft were drawing to a close.

The MB 5

James was fifty three but it was not the end of the Martin-Baker saga, only the end of the beginning and his greatest achievement was yet to come. In April 1944 there were discussions within the Ministry of Aircraft Production of possible methods of emergency escape from high speed aircraft. By August it had been concluded that the best way forward was the forced ejection of the pilot probably in his seat. Two companies were approached, including Martin-Baker, for their initial thoughts.

By the end of the year James, never a slouch, had designed an ejector seat for future aircraft. The pilot and his seat would be ejected by an explosive, propulsive, charge and he would then detach himself from the seat and pull the ripcord of his personal parachute.

He soon recognised that the crucial matter was what accelerative force could be withstood by a human being? At that stage he met with a surgeon who explained the mechanics of the spine. The first tests on his ejector seat were carried out on a steeply inclined test track. The first test with a dummy was on the 20 June 1945 and with a man in the seat on the 24 June with no ill effect. But, after a test when an individual received a back injury, James realised that a steady acceleration generated an impact load at the instant of firing of this charge. The problem was not so much the acceleration but the rate at which it was achieved. He also recognised it was important that the vertebrae were set square to one another which depended on the posture of the man in his seat, and his head being secured to the head rest.

After further tests on a 65 ft inclined test track Martin-Baker were provided with a Meteor III aircraft to modify for flight tests. But in March 1946 James was very seriously injured in a contretemps with drunken gypsies making a considerable disturbance on land adjoining his home, Southlands Manor. He was fortunate to survive this brutal attack and it was six weeks before he returned to the factory against his doctors advice. At this junction Francis announced that he wanted to withdraw his capital. Again James's brother-in-law stepped in and provided the funds needed.

The modification of the Meteor was completed and, on the 8 June 1846, a dummy was successfully ejected from the aircraft sitting on the ground. On the 31 June a dummy was ejected at 2000 ft and a speed of 410 mph but some problems were experienced. After further modifications the first man was successfully ejected on the 24 July at 8000 ft and 310 mph. Much more testing and development was needed before, on the 15 December 1947, the Mark 2 seat was finally certified. On the 2 January 1948 it was approved for the Meteor 4 aircraft.

This was the beginning of a long line of seats developed for a large number of aircraft from many countries, not least of all America. Seats were developed for high altitude, escape from ground level and from supersonic aircraft. Rocket propulsion of seats was developed as an alternative to explosive propulsive charges. By 1951 Martin-Baker were employing about 400 and the factory had expanded by an additional 20000 sq ft. In 1954 James's sister set

The Mark 7 ejector seat.

up a factory in the Island of Man and in 1956 a subsidiary of Martin Baker was set up in Canada. Finally, in 1957, his sister bought a disused US Air Force Station at Langford Lodge on the shores of Lough Neagh in Northern Ireland and Denis Burrell, James's nephew, moved to set up a factory which remains in operation to this day.

A Rocket Propelled Ejector Seat with Dummy being fired from an AMTX Cockpit.

James Martin died age 87 on the 8 January 1981 having never managed to completely retire. By that time 4,788 lives had been saved by ejection from aircraft for one reason or another. At the time of writing over 7,000 lives have been saved. He was awarded an OBE 1950, CBE 1957 and a knighthood 1965. He was a FIMechE and Hon. FRAeS and had been awarded the RAeS Wakefield Gold Medal 1952, the Barbour Air Safety Award 1958, Cumberland Air Safety Trophy 1959 and the Royal Aero Club Gold Medal 1968. The Manchester Institute of Science and Technology made him an Honorary Fellow 1968, and the Queen's University of Belfast awarded him an Honorary DSc 1968.

James married Muriel Holmes on the 28 February 1942. She was a nurse he had met when he was in hospital in 1936. Despite his having been a workaholic all his life their marriage was a happy one, and they had twin sons and two daughters.

Sources:

1. 'Sir James Martin' by Sarah Sharman, pub. Patrick Stephens Limited, 1996, ISBN 1 85260 551 0.

2. 'Safety Seats' by R. Martin, Professional Engineering, Vol. 18, No. 16, Aug. 2003.

ERIC CHRISTOPHER STANLEY MEGAW

Eric Christopher Stanley Megaw was born in 1908 in the Old Portobello Hotel in Dublin which had been converted into a hospital. He was the first child of Arthur Stanley Megaw and his wife Annie. His father, a solicitor, was also a distinguished literary figure providing Eric with a childhood steeped in literature and music.

From an early age Eric had a precocious interest in finding out how the physical world worked. In 1916 he attended Mourne Grange Preparatory School in Kilkeel where he did not shine academically but began a lifetime interest in wireless. When he was 14 years old he moved to Campbell College, Belfast, where he developed a strong connection with Amateur Radio. His parents encouraged this providing him with a workshop at home in which he carried out radio experiments.

This activity was not just 'schoolboy meddling' and he contributed articles to the T and R Bulletin (the journal of the Radio Society of Great Britain) on his results. While at Campbell College he constructed a radio receiver and received the first verified Irish contact with New Zealand. This was, in fact, illegal and as such was not allowed by the school. This caused problems at school, when his proud father reported the fact to a local newspaper, which published details of the radio contact! At the same time, Eric expanded his love for music and the arts which he shared with one of his teachers at school who also happened to be his physics teacher.

In 1924 he was recommended by the school to take part in a signalling course in England, organised by the War Office. He received an excellent report in which his practical abilities and experience were praised. The report also pointed out, in particular, the benefit which Eric had obtained from the theory part of the course. In 1924 Eric produced the first transmitted amateur radio signal from Ireland and, in the next year, he recorded the first contact between Ireland and Australia which was a very significant achievement for a schoolboy. He also obtained patent protection for his 'improved vernier tuning of condensers' invention.

He enrolled in The Queens University of Belfast in 1925, to study Electrical Engineering. At Queens, he embarked on a lifelong friendship with T Palmer Allen, a young lecturer in Radio Engineering, who subsequently became Professor of Light Electrical Engineering in the University. 'TP' advised and encouraged Eric during his career. During the vacations in his undergraduate years, Eric carried out a series of experiments which contributed materially to the existing knowledge of the propagation characteristics of short wavelength (SW) radio waves. In his second year, he co-operated with the Royal Geographical Society to monitor the progress of McMillan's Arctic Expedition. In the same year, he installed a radio station on the ship 'Lord Antrim', to study the effectiveness of SW ship-to-shore radio, as the ship sailed to Canada. These results were reported in the T& R Bulletin of December 1926 and contributed to the validation of SW radio for ship-to-shore communication which was remarkable for a second year undergraduate.

Later in 1926, he initiated studies of SW reception between Belfast and a station in Rawlpindi operated by a friend and collaborator, Major Coates. He found consistent anomalies in propagation which required further study and explanation. At this stage in his undergraduate career, he developed an aptitude for and an interest in the complex mathematics underpinning the theory of radio wave propagation which was an absolute necessity for further progress in his chosen field. Eric's late 'conversion' should be an encouragement to those students who initially find mathematics difficult.

After graduating from Queen's with very high marks in Radio Engineering in 1928, Eric was awarded a Beit Fellowship to pursue research at Imperial College, London, under Professor Fortescue. His topic was 'improvements in thermionic valves'. Thus, Eric's experience of short wave radio propagation combined with his research into thermionic valve operation, provided the ideal preparation for his lifelong interest in the generation and application of very short wavelength radio waves. After his first year of research study, he obtained the Diploma of Imperial College (DIC) and, shortly after, was recommended to GEC Research Laboratories, Wembley, by Professor Fortescue. He joined GEC in 1930, to begin a fruitful association which was to last for 14 years.

On 29 March 1933, Eric married the Dutch girl who was to be a loving support until his death in 1956. They had two sons, the first born in 1934 and the second in 1941.

During the early 1930s, Eric worked on a new thermionic valve called the Magnetron which showed some promise as a source of microwave signals. In 1933 he received the IEE Duddell Premium for a paper on the Magnetron Short Wave oscillator. He continued his interest in radio wave propagation and in 1935, while assisting at a Royal Institution lecture by Lord Rutherford, he produced a demonstration, illustrating the principles of radio scattering due to free electrons in the upper atmosphere,which was praised by the great man. He became involved in propagation studies at metric, and centimetric, wavelengths over land and sea.

At the onset of war, the urgent need for high-power, airborne, radar equipment led Eric into the adventure for which he is normally remembered which is the development of an operational, lightweight, magnetron valve. GEC had already worked on this project since 1932. In 1939, the Research Group including Eric had produced a high power continuous wave (CW) valve at metric wavelengths and a low power CW valve at a wavelength of 3cms. After a visit by Watson-Watt - one of the pioneers of RADAR - to Wembley, work was initiated on a very short wave pulsed magnetron for Aircraft Interception (AI) Radar, leading to the development of a 500 watt valve operating at a wavelength of 10cms, in March 1940.

At the same time a team of physicists under Professor Oliphant at the University of Birmingham were also working on magnetron valves. J.T. Randall and H.A.H.Boot invented a new structure which employed 6 slot-coupled cavity resonators symmetrically surrounding an axial wire cathode. The resonators, which were milled out of a solid bar of copper, also acted as anode and heatsink. The provision of glass to metal seals at each end of the axial cathode allowed DC voltage to be applied. Since the glass to metal seals were only waxed onto the copper plates, and were leaky, continuous evacuation was necessary. The microwave output was taken from a wire loop in one of the resonators. The estimated CW output power at 10cms was 200-300 watts which was about ten times the CW power produced by any other group. This was very much a 'proof of idea' valve but, since it had to be continuously evacuated and required a large electromagnet weighing 50lbs and used a tungsten wire cathode which limited its power output, it was not a practical device outside the laboratory. However, it represented a major step forward in magnetron development.

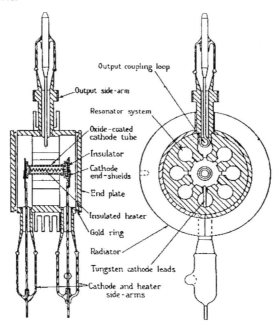

During a visit by the Wembley group to see the work at Birmingham, Eric was able to suggest improvements and it was agreed that further work would be collaborative, in order to make use of the existing body of practical design experience at Wembley. The need was for a compact, lightweight pulsed transmitter at 10cm for airborne radar, so Eric Megaw completely re-engineered the Randall and Boot design, turning it from a laboratory demonstrator, into an efficient product, that 'could be produced in quantity and requiring relatively unskilled labour'.

As the diagram shows, the valve comprises a cylindrical central electrode (cathode) surrounded by a concentric cylindrical second electrode (anode). A DC voltage is developed between cathode and anode. Electrons are released by heating the cathode. The combination of the radial electrical field due to the DC voltage, and an axial magnetic field, produced by a permanent magnet, causes the electrons to rotate in the space between cathode and anode. The path traced by the electrons depends on their velocity; fast electrons follow a tight circular path, while slow electrons follow a wider path and eventually reach the anode.

The anode contains eight evenly distributed resonant holes or cavities. As the electrons pass the mouth of the cavities the majority give energy to the cavities, slow down, and eventually reach the anode. Some electrons receive energy from the cavities, accelerate, and spiral back to be absorbed in the cathode.

If the spacing and resonant frequency of the cavities are appropriately related to the rotational velocity of the electron beam, net energy is transferred from the electron beam (and hence from the DC fields producing it), to the radio frequency fields developed in the cavities. A loop inserted in one of the cavities is used to extract microwave power.

Eric Megaw created a completely sealed resonant cavity structure using a thin shim of gold between the copper plates which diffused into the copper on being heated. As a result of co-operation with SFR Laboratories, Paris, he replaced the wire cathode by a large oxide-coated internally heated cathode . He reduced the length of the axial cathode sufficiently to enable a standard 6-lbs magnet to be used. This re-design made use of the 'rules-of-thumb' which derived from the previous work at GEC.

A whole series of magnetron valves resulted and in July 1940, a Model Number E1189 produced 3kW at 10cm. The rate of progress from discussions at Birmingham in April, to a re-engineered operational device in July, gives some indication of the ingenuity of the scientists and of the pressure under which they were working. Eric's group produced E1189 No12, which was a revised design with 8 resonators, a larger cathode and a longer operational life and was the valve taken by Sir Henry Tizard and his mission to MIT Labs in Boston, in late 1940, as part of the collaboration between USA and GB. The Americans were very impressed because at this time they only had valves called klystrons which gave 10Watts at 10cms.

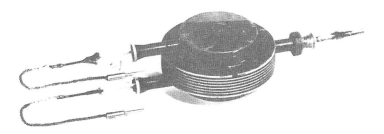

The E1189 magnatron.

A Short's Stirling fitted with airborne radar at Sydenham Airport, Belfast.

Eric Megaw's contribution was central to the eventual success in counteracting the threat from German bomber raids. In 1941 he was summoned to Buckingham Palace to receive the MBE from the King, in recognition of his contribution to magnetron development and thus to the Battle of Britain. Sir Edward Appleton noted, 'Those who were in the business, knew well how much the practical development of the cavity magnetron for operational use was due to Dr Megaw. Yet, smilingly, he let the credit go wholly elsewhere'.

After the magnetron interlude, Eric maintained his interest in the propagation of Short Waves. He published a paper in 1946, summarising the experimental studies on Short Wave propagation carried out during the war for which he was awarded an IEE Premium.

In January 1946, he left GEC and joined the Royal Naval Scientific Service, where he was promoted to Chief Scientific Officer and in1950 he was appointed Director of Physical Research. He was still passionately interested in propagation research and drove himself hard to combine research with this demanding administrative role. In 1946 he received a DSc from Queen's and in 1947 he was elected MIEE, by the Institution of Electrical Engineers. Having for many years contributed to the affairs of the IEE, he became Chairman of the Radio Section of that Institution , where 'He did much to stimulate and encourage Radio Engineering'. His Chairman's address in 1953 was on the topic of 'Radio Propagation and Scattering'.

On the evening of 24 January 1956, Eric was working on a major paper, in which he hoped to clarify the theoretical underpinning of radio-scatter propagation, a precursor to tropospheric-scatter propagation, which is now used for 'over the horizon' communications.

Later that evening, his wife went to bed, expecting him to follow when he had completed his work. Tragically, in the morning she found him in bed, apparently sleeping but he was dead. In 1948 Eric had an attack of rheumatic fever from which he made a very slow recovery and this, together with overwork, may have contributed to his death at the early age of 48years.

The paper he was working on was completed by two of his colleagues, F.A. Kitchen and M.A. Johnston, using Eric's notes and was published in Proc IEE. Mr Kitchen wrote "Dr Megaw has made a further important step towards the complete understanding of the mechanisms of tropospheric forward scatter". The paper was posthumously awarded their highest Premium by the IEE.

This brief story of a truly great Ulsterman does not do justice to his character, which may be better inferred from comments on his untimely death.

Sir Edward Appleton one of the pre-eminent scientists of the time who gave his name to the atmospheric Appleton Layer in a letter to a member of the Admiralty staff wrote, "It is, of course, a great thing to be successful, and to inspire universal esteem is something much greater and that certainly describes Eric, but then one must mention the graceful way he lived his life and the way he maintained the best standards. Just anything would not do. The loss is now drenching and there seems no other word for all of us. British science can ill-afford to lose someone so able and so human and so gay with it all".

The Chief of the Royal Naval Scientific Service stated that "Megaw was not only gifted as a scientist, he possessed personal qualities which contributed no less to the successful outcome of his endeavours. His integrity, sincerity, loyalty and friendliness, combined with a high sense of duty, made him an ideal leader. He will be sadly missed".

Sources:

1. 'A Backroom Boy' by Arthur Stanley , pub. W Erskine Mayne Ltd. 1960

2. 'Technical History of the Beginnings of Radar,' by S.S.Swords, pub. IEE

3. 'Radar Days', by E G Bowen, pub. Adam Hilger 1987

4. 'A Radar History of World War 11,' by Louis Brown, pub. IOP Publishing 1999

5. 'Metres to Microwaves', by E.B. Callick, pub. Peter Peregrinus 1990

ALEXANDER MITCHELL

Alexander Mitchell was born on the 13 April 1780 in Dublin the eighth child of William Mitchell, born in Dublin, who was Inspector-General of the British Barracks in Ireland. When Alexander was seven the family moved to Pine Hill near to Belfast. On the death of his father in 1790 his mother moved with her remaining daughter and three youngest sons to a rented cottage named Eglinby Cottage a mile from Belfast.

Up till the age of eight or nine, when Mitchell went to school, he was educated at home by his mother. When he was ten or eleven he attended the best classical school in Belfast at that time – the Old Academy in Donegall Street. While still at school his eyesight began to fail, probably due to amaurosis. This is a disease which leads to paralysis of the eye without any observable deterioration of the eye, so people do not realise that the sufferer is blind. By the time he was sixteen or seventeen he could no longer read, and in his twenty first year he was forced to give up letter writing.

At school he showed a remarkable liking and aptitude for mathematics, in a curriculum devoted mainly to classical studies, as was the norm for that day and age. By the time he was sixteen he had acquired a great deal of learning which was to stand him in good stead for the rest of his life. This formed the foundation on which he extended his knowledge by conversation and being read to by others. His family recognised him as a font of knowledge over a wide range of academic subjects.

After he finished school in 1796 there is little recorded until 1801 when, despite his mother's strong disapproval, he married Mary Banks the daughter of a near neighbour. They settled in a two storeyed cottage situated between the Newtownards and the Albert Bridge Roads on the site where the Ropeworks was later to be built, where they brought up their five children.

To supplement his very slender patrimony he started up the manufacture of bricks, and also built twenty houses in and around Belfast. For the next thirty years this provided a modest income which allowed him to indulge in his various interests including mechanical experiments. It was not until 1832, when he was in his early fifties, that a friend asked his advice on the construction of a cheap facility to dry dock ships of moderate size for repair. This led to Mitchell's patent of 1833 – 'A Dock of Improved Construction to Facilitate the Repairing, Building or Retaining of Ships and other Floating Bodies'. His address was given as Brickfield in the Parish of Ballymacarrett, County Down and he is referred to as a Civil Engineer.

The patent describes what are basically wooden watertight pontoons, or caisons, which are connected together with massive longitudinal beams to form the floor of the dock. Each of the pontoons are restrained by four piles of posts reinforced with wrought iron plates. Large wrought-iron pins or pegs can be fitted in transverse holes bored through the wrought iron plates and pile. At low tide pins are inserted which restrain the pontoons from rising with the tide. At high tide the vessel to be dry docked is floated in above the pontoons. At the next low tide the vessel comes to rest on the pontoons, and all the pins are removed. As the tide rises the pontoons lift the vessel and at high tide the pins are inserted in holes higher up the piles or posts, to prevent the pontoons moving down as the tide recedes. What he proposed was basically a hydraulic ship lift.

The most important part of the patent relates to the piles or posts. He recognised that these can be driven into the ground by a pile-driver, as is conventional practice to this day. But he also envisaged a new method of screwing the pile into the ground by applying a torque about the axis of the pile. A cast iron screw with about one and half turns is fitted over the end of the wooden or wrought iron pile, and attached to it by iron pins with their ends riveted. The end of the pile has an iron pointed ferrule to allow the pile to more readily penetrate the ground.

Screw Pile

It is clear from his 1848 paper in the Institution of Civil Engineers, Mitchell recognised in his dry dock design that there was a considerable upward force on the pile when the pontoons were held down against a rising tide. The helical screw, which it was envisaged might be up to five foot diameter, provided a great resistance to the pile being pulled out as well a considerable load bearing area for downward loads.

Prior to the patent in 1832 Mitchell accompanied by his son John, then aged nineteen, carried out a test on a model screw pile. One evening they hired a boat and rowed out to a sand bank in Belfast Lough, where they screwed down the model pile, leaving it standing proud of the water. Next day they returned and found it still firmly fixed.

Subsequently, before undertaking any work using a big screw-pile, Mitchell used a model test to determine the suitability of the ground and its load bearing capacity. A wrought iron rod of one and a quarter inch diameter with a six inch diameter screw was screwed down 27 feet or as far as it would go. It was then loaded by the weight of 12 men – approximately 1 ton – standing on a platform resting on the capstan firmly attached to the rod whose primary purpose was to apply the torque to the shaft. If it did not sink perceptibility relative to a second unloaded rod then the size of the helical screw to support the design load could be determined. For example a four foot diameter screw could support at least 64 tons.

Though the patented dock was never built screw piles were extensively used world wide mostly, but not exclusively, for maritime purposes such as moorings, harbour works, breakwaters and more especially lighthouse structures. In 1838 Mitchell laid the foundation of a lighthouse on the Maplin Sands off the Essex Coast. Nine five inch diameter wrought iron piles with four foot diameter cast iron screws were screwed down to a depth of 22 feet. A 36 foot square raft resting on the sands at low tide, provided a platform to accommodate the forty men needed to turn the capstan to screw the piles down. The piles remained untouched for two years to check if any change occurred.

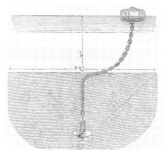

Screw Mooring

In the meanwhile he completed a lighthouse in Morecombe Bay off the entrance to Fleetwood Harbour. This was supported on seven 16 foot wrought iron piles with three foot diameter cast iron screws. Six of these formed a hexagon and the seventh, at the centre, basically supported the lantern. The bottom ends of seven 14 inch square timber bulks were bored out with a 5 inch diameter hole extending for 7 feet, to fit over the exposed ends of the wrought iron piles. They were reinforced with iron hoops driven on hot.

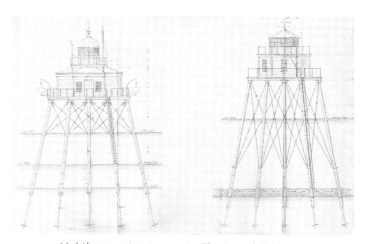

Lighthouse at entrance to Fleetwood Harbour.

In the following two decades Mitchell was responsible for building many lighthouses round the English and Irish coasts. These included one off Carrickfergus in Belfast Lough which served both as a lighthouse and a pilot station. He also used screw piles to construct piers such as at Courtown, Wexford, and breakwaters such as the Government breakwater at Portland on the Dorset coast. In India screw piles were used for the viaducts and bridges on the Bombay and Beruda railway. The screw pile principle was also adopted to moorings for ships, which was covered by the 1833 patent. The screw pile principle was even adopted to a screw tent peg, which would be a blessing for modern day campers.

Screw Tent Peg

In the United States the first screw pile lighthouse was built in 1843 at Black Rock Harbour in Connecticut. This was designed by Captain William Swift who had travelled to England to inspect the Maplin Sands Lighthouse and consult with Mitchell. Subsequently he built the lighthouse on the Minots Ledge off Cohasset, Massachusetts, which in 1851 only a year after its completion was destroyed in a severe storm, drowning its two keepers. The cause was ascribed to inadequate cross bracing between the legs. Many other lighthouses using screwed piles were built along the East Coast of America including the massive 46 foot lighthouse on the Brandywine Shoal in Delaware Bay, which had 32 screw piles. This was designed by Major Hartman Bache and Lieutenant George Meade in conjunction with Mitchell who visited Delaware in his seventieth year.

Besides the screw pile Mitchell, in 1853, turned his attention to improvements in screw propellers, which he patented. He claimed in his patent that his design gave a higher propulsive efficiency and reduced the vibration in the ship and cargo. The propeller was first tried on Erin's Queen and the Mulvina, and their captains claimed that it gave them increased speed and reduced fuel consumption. At a meeting of the Screw Steamship Company James Hamilton complimented Mitchell on his design of propellers.

The 'Seven Foot Knoll' screwpile lighthouse built in 1856 and removed to a site on land at Baltimore, USA, in 1988 as a historic monument.

Mitchell had many interests outside of engineering. He was a member of the Belfast Natural History and Philosophical Society and in 1828 he read a paper entitled 'Meteoric Stones' and in 1833 a further paper on 'Runic Inscriptions in Scandanavia'. He was very sociable and entertained widely. When he lived in Cork temporarily during the erection of a lighthouse off Queenstown, he entertained and got to know Professor George Boole, the great mathematician, who developed Boolean algebra. He had a great love and understanding of music and he played the flute and accordion as well as singing Irish songs. At 6 foot he was a tall and friendly man who did not give the appearance of being blind.

He became an Associate Member of the Institution of Civil Engineers in 1837. On the 22 February 1848 Mitchell read his paper – 'Submarine Foundations' at the Institution of Civil Engineers in London. Later in that year he was elected a Member of the Institution and awarded its prestigious Telford Medal. In the Great Paris Exhibition of 1855, a successor to the 1851 Great Exhibition in London, Mitchell was awarded a Silver Medal for his screw mooring which was covered in the 1833 patent.

Alexander Mitchell overcame the enormous disadvantage of blindness to become known as 'The Great Blind Engineer of Belfast'.

His wife Mary who had been his mainstay died in 1864 at the age of eighty six, and he lived on to die four years later on the 25 June 1868 at the age of eighty eight.

Sources:

1. 'Alexander Mitchell, the Famous Blind Engineer of Belfast' by F.J. Bigger, Selections from 150 years of Proceedings, 1831-1981, pub. Belfast Natural History and Philosophical Society, pp 148-158, 1981.

2. 'Screwpile Lighthouses – Sentinels Secured to the Seabed' by Elinor De Wire, pub. Mariners Weather Log, Fall 1995.

3. 'Docks for the Repair, etc of Vessels'. British Patent No. 6446, granted to Alexander Mitchell 1833 and renewed for a further 14 years in 1847.

4. 'Submarine Foundations' by Alexander Mitchell, Proc. Inst. of Civil Engineers, vol. 7, pp 108-132, 1848.

Sponsored by Uprite Structures and Services

REX McCANDLESS

Rex McCandless was born on 2 May 1915 at Culcavey near Hillsborough, Co. Down, elder son of Joseph and Sadie McCandless (nee Cromie). His father was a farmer who worked some 100 acres of fairly rich alluvial land at Ballyroney. Unfortunately he also tried his hand at the stock market and lost everything in 1928. Joe emigrated to Canada to try and make his fortune leaving his wife to raise the two boys. This meant that Rex, having the responsibility of supporting his mother and brother, left school at the age of thirteen with no formal qualifications.

He took some special tuition and passed the entrance exams for a boy apprentice at RAF Halton. However, because of flat feet, he failed the medical and returned to his family in Northern Ireland where he obtained a job with the flour miller, E. T. Green of Belfast. To help him travel to and from work, an aunt gave him a 1923 Raleigh motorcycle. This awakened an interest in engines and gave him some practical experience in looking after it.

Rex moved back to England to work in the NAFFI first at RAF Uxbridge then at other bases before getting a job in the maintenance section of the Daily Herald transport department. With this experience he returned to Northern Ireland to start work in 1939 in the newly established Shorts Factory initially installing the braking system in Bristol Bombay bombers which were the first product of the Queen's Island factory. The security of the job led him to upgrade his motor cycle from a fairly basic 250cc Triumph to a new 1940 Triumph Tiger 100 as some new motor cycles were still available despite wartime restrictions.

There he also made friends with Artie Bell who, before the war, had established a motor cycle sales and service business but was doing his bit for the war effort in Shorts. Artie was a keen motor cycle racer and had a modified Tiger 100 which should have been faster than Rex's standard model. He introduced Rex to road racing and the pair took part in the 1940 Irish 500 Road Race Championship held in Phoenix Park, Dublin. The field included some experienced riders, such as Ernie Lyons and Terry Hill as well as Artie Bell, yet the novice Rex won taking fastest lap as well. This showed that what Rex described as "fiddling" with his engine had a positive effect and gained him a growing reputation as a mechanic.

This encouraged Artie Bell to re-open his Woodstock Road premises and undertake sub-contract work for the Ministry of Supply in areas such as repair of cycles, motor cycles and light transport vehicles. Although he had a share in the profits from the business, McCandless decided to strike out on his own and left Bell on good terms. He set up in a yard off Dublin Road which had been stables and quickly established a good reputation as a general repair man. A typical task was to convert an articulated lorry into a rigid one. This was achieved by welding the rear section of a wrecked bus chassis in place of that of the tractor unit. With considerable activity in airfield construction around Belfast, there was a requirement for servicing and repair of earth moving equipment. McCandless moved to fill this gap and, as business expanded, took his brother Cromie, who as well as having some practical engineering knowledge had experience of office work, into partnership with him. The ongoing success of the business allowed McCandless the freedom to follow his interests in other directions.

Further expansion required larger premises and these were supplied by Artie Bell in return for an equal share in the partnership. This work provided the three with a secure base from which to continue their motor cycling lifestyle. In particular, Rex continued to experiment with his Triumph Tiger in an attempt to improve its handling because he felt that, under certain conditions, it was in charge of him, a feeling he did not like. His instincts told him that no matter how strong a single plane frame might be resisting vertical loads, once it was subject to out of plane forces generated when cornering and deflections due to imperfections in the road surface, it could bend laterally. He also realised that the wheels were not always in contact with the road surface. As they bounced there were rapid changes in the forces being applied to the frame. This meant that the front and rear wheels took their own lines through a corner and these were not necessarily the line intended by the rider. Combined, these problems led to unsatisfactory handling. He decided that soft springing and good damping were the key solutions while, at the same time, providing a frame which was more rigid in all three planes.

Current motor cycle design called for relatively short spring travel for the front wheel and none for the rear. McCandless started by adding a sprung rear end to a standard single plane frame then developed his ideas by building himself a light, double loop, frame on which almost everything was adjustable. It was in practice a mobile test bed for his experiments. When motor cycle racing started again after World War Two, he used the new frame with the Triumph Tiger engine to good effect in the Bangor Castle races. The rear suspension improvements were demonstrated in 1946 on the single plane frame when a challenge match was arranged on a grass track circuit at Brands Hatch in Kent between an English team and an Ulster one. The English riders were the top competitors of the period and the Ulstermen were virtually unknown. The sceptics looked askance at the ungainly rear suspension of the Ulstermen in the sure knowledge that rigid rear ends were essential to win on grass. When the Ulster team swept the boards, there was a great clamour to get copies of the system. McCandless, who was basically a development engineer, and who never liked the thought of having to repeat operations, such as were required in manufacturing a product, sold the rights and went on to the next problem.

At a motor cycle race meeting in 1945 McCandless met the legendary Freddie Dixon who, after a successful career on two wheels, had transferred to four and was a legend from his exploits at Brooklands and the Ards TT circuit. Dixon was trying to develop some ideas on four wheel drive and needed assistance but, because of work permit regulations, was having difficulty in obtaining the sort of person he needed. A large part of Dixon's reputation was built on his skill with carburettion. While working with Dixon in an unpaid capacity, much of this skill was passed on to McCandless who put it to good use tuning the motor cycles of his friends. A typical example was that of Ernie Lyons, the experienced Triumph works rider who benefited from McCandless's help at the 1946 Manx Grand Prix. His machine had been poorly tuned for the Ulster Grand Prix in August but application of the wizardry of McCandless and Dixon led to a win in September, despite a broken frame.

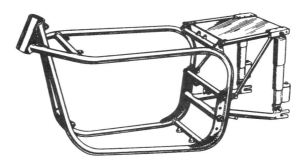

The "Featherbed" Frame

In 1949 McCandless managed to convince the management of the Norton motor cycle company that their approach to racing by continuing to use their old frames and making them heavier was wrong and that results would be improved by using his duplex frame. He was proved right and, by using the "featherbed" frame, Nortons were able to compete internationally and remain at the top for several years without investing in new engine technology. The "featherbed" frame made its debut in 1950 at a meeting at Blanford Camp in Dorset and then took first three places in both Senior and Junior TT races in the Isle of Man with Artie Bell first in each class.

The international competition was fierce and the domination given to Norton by the superb handling of the duplex frame did not last long. McCandless, who only acted as a consultant to Nortons, disagreed with their ideas and produced some new versions for demonstration purposes. The most successful was his "kneeler" design in which the duplex frame top rails sloped from the steering head to meet the lower ones just above the swinging arm pivot. The rider adopted a prone position supported by pannier fuel tanks and the whole machine lay within a streamlined shell. The improved aerodynamics and lower centre of gravity combined to produce a superb racer. Ridden by Ray Amm within two weeks of completion, it achieved the fastest lap at the North West 200 race in 1953, but failed to finish. It later achieved many world speed records at Montlhery including the 500cc one hour record of 133.66 miles which had previously been held by a Gilera at 124 miles. The "kneeler" design formed the basis of Eric Oliver's very successful World Championship sidecar outfit and was widely copied.

In 1953 McCandless stopped working with Nortons and returned to Belfast. While with Freddie Dixon, he had been involved with a four wheel drive vehicle project the aim of which had been to try for the world land speed record. In the interval, Harry Ferguson had bought the project, renamed it Harry Ferguson Research Ltd., and changed the emphasis to designing a four wheel drive motor car one version of which was to be a light cross country vehicle to carry four soldiers and their full kit. McCandless approached Ferguson with his ideas of how to build such a vehicle and the latter agreed to fund it. Again, light and strong were the guiding design principles. By combining a rear-mounted motor cycle engine and chain drive through a backbone chassis which doubled as an oil-bath, McCandless came up with a design he called "the Mule". By incorporating a differential gear for each axle, arranging for the front wheels to be driven three percent faster than the rear and incorporating a freewheel unit which locked up if one wheel started to slip he solved the problem of supplying power to wheels as and when required. By using a diamond seat pattern the vehicle would be evenly loaded with any number of passengers from one to four. The machine achieved all its design objectives and was demonstrated to military top brass who received it enthusiastically.

Chassis of the first cross-country 'Mule'

For a variety of reasons it never went into production and was the cause of a major falling out between McCandless and Ferguson. However at the same time as developing the "Mule" McCandless built a similar, but single seat, machine for the race track and won regularly against competition with up to seven times his engine size. He then went on and developed, in 1956, a sports car and a trials car using the easily obtained Ford 1172 cc side valve engine. Both these designs incorporated the engine as part of the backbone

Rex with the second 'Mule'

chassis design and were very competitive. They could have easily been developed into production versions but McCandless never found production interesting. As Ford introduced a new series of cars in 1957 with overhead valve engines and stopped making the side valve one, the project was abandoned.

Laurie McGladdery, who had shared the driving of the four wheel drive racing car was a manufacture of bricks in Belfast. He offered the McCandless brothers a new site for their workshop on a worked out clay pit alongside his Limestone Road premises. He was having

McCandless Sports Car made for Dorothy McGladdery

problems with the quality control of the brick making process. This involves a gradual heating of the clay to drive moisture from the inside of the brick then using a high heat to fuse the material. McCandless observed the inefficiency of the combustion process as coal was shovelled into the top of the kiln and also the variation of draught from a tall chimney stack. By providing the fuel as pulverised coal dust and delivering it with a predetermined airflow to the parts of the kiln where it was most beneficial, McCandless both improved the quality of the end product and reduced costs. Again he sold the manufacturing rights of his design and moved on to other problems.

As the repair business continued to prosper, the McCandless brothers bought a de Haviland Puss Moth in 1959. Even this light aircraft required a considerable runway from which to operate and therefore limited its flexibility and use. This set McCandless thinking about aircraft which required shorter take off runs and led him to look at the Autogyro. He built himself a small one using a Triumph motor cycle engine. Later versions used a modified Volkswagen engine. At the time the Autogyro suffered a bad press as there had been several fatal accidents, one in front of large crowds at the Farnborough Air Show. McCandless built a test rig which included some very large concrete blocks to prevent it from becoming airborne. With this he was able to demonstrate how the main rotor could stall in certain flight conditions which coincided with the Air Show crash. He believed that his lack of formal educational qualifications meant that the authorities did not even give his ideas a second glance. However, after battling for about

First Autogyro

Later Autogyro with Volkswagen Engine

twelve years, the Air Registration Board accepted that his analysis of the problem was correct and certain manoeuvres, particularly swoop and climb, were made illegal for Autogyros. Again, although he had built some six variants of his Autogyro design, he sold the rights to Dr Bill Ekin of Crumlin who attempted to develop it into a commercially viable project. This was unfortunately impractical due to serious blade vibration problems.

Rex McCandless died unmarried on 8 June 1992. In retirement, he lived quietly in Killough with Dorothy McGladdery pottering with various inventions and occasionally regaling local Technical College students with his life story. He believed that these talks could encourage the students to adopt his approach to life and, by looking critically at any problem and querying accepted wisdom, find an innovative solution.

Sources:

5   'Motor Makers in Ireland', by John S. Moore, pub. Blackstaff Press, Belfast, 1982.

6   'Sweet Dreams', by Gordon Small, pub. Ulster Folk and Transport Museum, Holywood, 1989

7   'To Make a Better Mousetrap', by R.L. Jennings, pub. Jennings Publications, Belfast, 2002.

WILLIAM JAMES PIRRIE
VISCOUNT PIRRIE OF BELFAST

Lord Pirrie has been described as "the best salesman of his time" in that he could sell a ship - in the form of a contract to build a ship - more skilfully or easily than any man of his generation. However, he was much more than that; he was a great planner long before the term had any economic significance. His skill lay in ensuring that his workmen had all the necessary materials of the right quality in the right place at the right time and for the right price. Throughout his life his main priority in any decision or course of action was to ensure the benefit and survival of Harland & Wolff.

William James Pirrie was born in Quebec on 24 May 1847. His father, James A Pirrie had been born in Belfast on 27 November 1821 and been sent to Canada to develop a timber business for his father Captain William Pirrie. James had married Eliza Swan Montgomery of Dundesart, Co. Antrim in Quebec on 28 June 1844. The Pirrie family was interconnected with the major shipping companies which operated in Belfast at the period. Captain Pirrie was a member of the Harbour Board who had pressed hard for the construction of the new channels into Belfast port which resulted in the formation of the bank to the east of the river Lagan first known as Dargan's Island then Queen's Island. Soon after James A. Pirrie's death in Canada in 1849, his widow returned to Northern Ireland to Captain Pirrie's home in Conlig.

Following his education at the Royal Belfast Academical Institution, W. J. Pirrie was enrolled as a gentleman apprentice at the shipbuilder, Harland and Wolff on 23 June 1862. He passed through the accepted route to management of apprentice, draughtsman, Assistant Manager, Sub-Manager and Works Manager. This gave him wide experience and a good overview of the company and its operations. He had followed Walter H. Wilson as apprentice and then as draughtsman. Pirrie spent some time learning about the design and construction of marine engines at the Greenock firms of McNab & Co. and Caird & Co., (which latter company was eventually purchased by H&W in 1916). He also took a considerable interest in the finances of the firm and the methods of financing the construction of new vessels.

In 1874, at the age of 27 he was promoted to a partnership. The other partners were E. J. Harland, G. W. Wolff and W. H. Wilson. His task was to gain orders and he did so with relish, interviewing any owner who was known to be considering building further ships. He realised that there were different requirements for different sea routes and a knowledge of the port facilities around the world was essential for advising customers. To improve his knowledge of the market, he travelled frequently both on his own ships and those of competitors and studied the facilities offered by the rising hotels around the world. This resulted in design developments and changes to the conventional layouts with an emphasis on the comfort of passengers, even third class ones. As was common practice at the time, and as a means of obtaining orders, the partners frequently invested capital in the shipping line either on their own or on behalf of the firm and therefore his personal wealth increased rapidly.

On 17 April 1879 William J. Pirrie married Margaret Montgomery Carlisle. She was his first cousin and some ten years his junior. Her elder brother, Alexander Montgomery Carlisle, served his time as a premium apprentice at Harland and Wolff and had been appointed shipyard manager late in 1878. At this time the firm constructed a new engine and boiler works alongside Queen's Road and laid out an expansion of the shipyard on the former pleasure gardens to the north of the existing yard.

Patent Steering Gear

In 1882, two patents were taken out by the four partners. They were for "Steering Gear" (no. 4637) and "Raising and Lowering Ship's Boats" (no. 5775). The principle of the former was to use springs as the spokes of a wheel attached to the rudder shaft so that the waves did not transmit sudden shocks to the rest of the steering mechanism. Over the next few years Pirrie and Wilson took out six more joint patents relating to various modifications in shipbuilding practice. Pirrie's last patent, in 1895 was for a steam winch which could handle more than one derrick at a time thus speeding the loading and unloading of cargo.

Shipyard in the late 1890's

By 1884, both Harland and Wolff had ceased to be active partners in the business leaving the day to day control to Pirrie. Wilson was the naval architect responsible for designing the ships which Pirrie built and sold. The firm officially became a limited liability company in December 1885 as 'The Queen's Island Shipbuilding and Engineering Company Limited'. The name was changed to Harland and Wolff Limited in 1888. At the same time the works were refitted with new machinery capable of handling the increased size of plates available from the iron founders who were now using open-hearth steel-making plant.

In June 1896, the yard suffered a serious fire which started in the woodworking area of the yard and destroyed a great part of the south yard and the adjacent engine works of Workman Clark. The need to rebuild quickly to fulfil the existing orders was tempered with the need to expand to be able to cope with the rapidly increasing size of vessels which were being ordered. As a result, the new no. 2 berth was built to the north side of the existing facilities capable of holding a ship of 700 feet by 70 feet. A cofferdam was constructed to enable work on the stern to be independent of the state of the tide. However, the main innovation of this berth was the gigantic gantry around a hundred feet high which carried mobile, hydraulically-operated, cranes for lifting the ever larger plates. They also carried the hydraulic riveting machines which were required to speed production generally and in particular deal with the double bottoms which were a standard feature of Harland & Wolff built vessels. It is believed that Pirrie, who so greatly admired the advances in production techniques being developed in America, based the system on the Newport Mews yard. Over the next few years, this gantry was extended to cover the adjacent berths and more gantries were built over other slips. Continuous improvements were being made to the port facilities by the Harbour Commissioners and by Harland & Wolff to their engine works. A new building, the Alexandra Docks Works, was constructed for the repair department which was growing as fast as the main business.

To cope with the cyclic nature of the shipping trade which mirrored the booms and depressions of the late nineteenth century, Pirrie had developed a secret commission club of specially favoured customers who had some of their new vessels built at no profit to Harland & Wolff or for cost plus a small, agreed commission. This combined with intermeshing investments between the ship-owners and the constructor ensured continuity of full order books and good levels of production which, in turn, justified and paid for the continuous improvements in the production facilities.

In 1894, Pirrie was elected to the Belfast City Council and the Board of the Belfast Harbour Commissioners. Also, following the death of Sir Edward Harland in December 1894, he became the chairman of the firm. In 1896 Pirrie was elected Lord Mayor of Belfast. His Liberal Unionist political opinion was shared by his wife but he renounced all political affiliation during this term of office and this offended some of the more hard-line unionists of the day. He even went as far as proposing measures to ensure Nationalist and Labour opinion had permanent representation on the Council. He was raised to the peerage as Baron Pirrie of Belfast in 1906.

The Hydraulic Riveting System

With the continued expansion of the business and the complications of his shipping interests, Pirrie felt it necessary to move his operational headquarters to London in 1898. This move strengthened his hold on the company as all his subordinates gave him their full information but only knew as much of other strands of the business as he wanted them to know. The expansion of the yards in Belfast kept them all busy. However, the end of the Boer War led to a slump in shipping and a further complication arose in international shipping when J. Pierpont Morgan, an American railway baron and steel magnate decided to expand into trans-Atlantic shipping. At that time, most of the Atlantic trade was dominated by European companies in various alliances including Pirrie's informal commission club. Morgan's strategy was to absorb and control as many of the existing groups as possible, paying for it with shares in his International Mercantile Marine (IMM). Seeing that such a group would have the power of life and death over shipbuilders, Pirrie made an alliance with Morgan which resulted in Harland & Wolff being given all new orders and repairs of the group. There was considerable opposition to the formation of IMM because it was thought that British interests would be subordinated to those of the USA and international finance. Others thought it had the potential to be just another share swindle. At one point it seemed that things were going so badly that these critics could have been right and IMM would crash. However, Pirrie fought on and had his nominee, J. Bruce Ismay, of the Oceanic Steam Navigation Company (and Owner of the White Star Line) made president of IMM. Between them they brought the conglomerate back to profitability. Pirrie managed to head off the American threat to Harland & Wolff without losing control of the company or the good relationship with those members of his commission club who were not in IMM.

Meanwhile other shipyards were forming ever larger groups and combining with the steel makers and forgemasters who supplied their raw materials. Pirrie saw that this could spell potential disaster in a depression. The shortage of skilled workers in Belfast led to expansion through the purchase of premises in Southampton. This required investment in building and plant. He was also investing in the new giant slips in Belfast and committed to further loans to IMM. He therefore sought a merger with John Brown & Co. This was quickly agreed and, despite selling 310 of the 600 authorised shares in the firm to Browns in May 1907, Pirrie remained as chairman of Harland & Wolff.

The benefits of his salesmanship continued and soon led to the need for further expansion. Following disagreement with the Belfast Harbour Commissioners over the lease of more land and the continuing lack of skilled men in the Belfast area, Harland & Wolff was forced to look elsewhere. As a result, in 1912, Pirrie bought the London and Glasgow yard in Govan and the Lancefield Engine Works formerly owned by David Napier & Co. The Govan yard had been recently modernised and rapidly became the centre for Harland & Wolff activities on the Clyde. At much the same time Pirrie also took over the Liverpool repair and maintenance facilities of the White Star and Leyland lines. These, however, required extension and modernisation before they could be profitably utilised. It was 1913 before H&W's Liverpool yard opened.

The engines for Harland & Wolff ships had progressed in line with the competition through compound to triple expansion. Initially they had been bought in as needed but over the years an engine shop had been built and design experience had been gathered. Steam turbines were developed by Charles Parsons and Co., and by the American company, Curtiss. Low-pressure turbines were developed to use the remaining energy in the exhaust from the triple expansion engines. The merger with John Brown brought with it a licence to use Parsons technology and the first use was in the White Star liner *Laurentic* in 1908. Later White Star liners *Olympic*, *Britanic* and *Titanic* also used this form of drive with the triple expansion engines powering the outer propellers and the turbine the central one.

Turbine engines, while smooth running, were not capable of providing improvements in efficiency. The heavy oil engine had been designed by Diesel and marine versions had been developed by his own firm MAN in Nuremberg, by Sulzer Brothers in Switzerland and by Burmeister and Wain of Copenhagen. Pirrie successfully negotiated with the latter for a licence to build their design and formed a new company, the Burmeister and Wain (Diesel System) Oil Engine Company which started operations in 1912 in the Lancefield Engine Works under the direction of O.E. Jorgenson.

During 1912, Pirrie took a small but active part in the Home Rule debate by arranging a public meeting at which Winston Churchill spoke in favour of the Bill being considered in Parliament. This led to some ill feeling against him, which he ignored. He also suffered an enlarged prostate gland and survived the necessary surgery but took some time away from his desk to recover. On his return, he feared that there might be a civil war in Ireland resulting in loss of production from H&W. He therefore made plans to increase the company's facilities on the Clyde by rebuilding the Govan yard, purchasing another yard and arranging sub-contract facilities with other shipbuilders. He wove a complicated system of finance through discounted bills from his customers the majority of whom were in the IMM or the Royal Mail Group. The financial difficulties of the IMM group placed enormous difficulties on H&W's resources but Pirrie was the only person with full access to the figures.

The declaration of war in August 1914 put H&W at a severe disadvantage because they had no Admiralty work on hand. Initially work concentrated on IMM vessels but by December some war work was available including the floating gun platforms known as 'Monitors' and conversion work from liners to troopships. This work relieved the pressure on the company's finances and allowed continued investment in new plant. H&W also purchased more Clyde shipyards which Pirrie intended to use for civil shipbuilding as soon as possible because there was a growing shortage due to the activities of the German U-boats. Orders for building new ships were organised under the new Ministry of Shipping. Despite a massive expansion of all the shipbuilders in the UK, the standard ship programme failed to meet its target and Pirrie was asked to become the Controller General of Merchant Shipping. This task he undertook with considerable gusto and by the end of the war had streamlined both design and production throughout the whole of the UK. As a result of the successes achieved in this post, Pirrie was created a Viscount in 1919.

After the war, Pirrie consolidated the order book of vessels agreed during the war but which were embargoed by the Ministry of Shipping. He continued a massive refitting and re-tooling exercise with the view of developing designs acceptable to the ship owners but nearly standardised. The new East yard in Belfast facing the Musgrave Channel was commissioned. By forming the British Mexican Petroleum company, he obtained a steady source of fuel oil for the diesel engined ships he proposed to make the mainstay of his production programme. To ensure continuity of supply of iron and steel, he had to take control of a steelmaker. Having been unsuccessful in a bid to take over David Colville & Sons of Motherwell in 1917, he reopened discussions and eventually bought the company in 1920.

At the same time he became increasingly worried about the future of Ireland and saw that Home Rule could have a disastrous effect on H&W. To this end he expanded the firm's interests on the Clyde by building the Clyde foundry to supply the Finneston Engine Works. When it was realised that Finneston could not supply engines for all the vessels being built by the company, and its subcontractors on the Clyde, H&W bought the Scotstoun factory of the Coventry Ordnance Works.

The freight and shipping markets collapsed in 1919 and 1920. The Government introduced measures to combat the overall recession and Pirrie jumped at the opportunity to have public funds pumped into the business to keep his continuous modernisation programme moving forward. It was not until 1922 that new orders were placed with H&W. During the lull the company developed the manufacture of land based diesel electric sets.

Pirrie suffered a fall in October 1923 which resulted in a dislocated left shoulder. He also suffered a recurrence of the prostate problem and his health started to deteriorate. He had planned a trip to South America to inspect new port facilities and shipping prospects. Although the business at home was not running as smoothly as he had expected, after a series of autocratic decisions, he went ahead and extended the trip to include the Pacific. There he caught a chill which developed into pneumonia and, although he was recovering when passing through the Panama Canal, he demanded to be taken on deck to see the sights. This had fatal results and he died on 7 June 1924 aged 77. His death showed up the flaws in his autocratic management as even his successor, Lord Kylsant, had no clear picture of the interconnecting, web like, structure involved.

Sources:

1. 'Viscount Pirrie of Belfast', by H Jefferson, pub. Mullan, Belfast 1948.

2. Belfast News-Letter

3. Belfast Telegraph

4. Northern Whig

CUTHBERT COULSON POUNDER

Cuthbert Coulson Pounder was born 12 May 1891 in Hartlepool, Co. Durham (later Cleveland), the fourth child among four sons and one daughter of John Pounder, master blacksmith, and his wife Margaret, daughter of Thomas Coulson, fisherman of Hartlepool. After the death of his mother, his father married (1900) Mrs Elizabeth Jane Smith (née Pounder) a distant relative and they had one daughter. His stepmother had a great influence on the young Pounder. He was brought up as a Baptist became a High Anglican and, later in life, a converted Spiritualist.

On leaving school he was apprenticed to the world renowned company of Richardson Westgarth and it was clear that he hated the workshops of his day but he was inspired by the design office under the leadership of L.D. Wingate. In 1916, aged twenty five, he joined Harland and Wolff as a draughtsman in the pipe arrangement office where, in due time, he became the Chief Draughtsman. During this period he was involved with the Holland-American liner, the *Statendam*, which was designed in 1920. Before it was launched in 1924 the market for North Atlantic liners had vanished and the hull lay untouched until 1927 when it was towed to Rotterdam for completion. All the machinery and the main pipe work was supplied by Harland and Wolff including the three stage Parson's reaction turbine which replaced the original Brown-Curtis turbines with a resulting improvement in fuel consumption of 40%. Pounder masterminded the changes in design and the transfer of installation information and the voluminous correspondence involved.

When Pounder joined Harland & Wolff steam propulsion dominated the marine world nearly exclusively. A fine example is provided by the three great ocean liners, the sister ships the *Olympic*, *Titanic* and *Britannic*, which had two immense reciprocating steam engines each of 15000 hp driving the wing shafts. These exhausted into a direct coupled Parsons' reaction turbine of 16000 hp driving the centre shaft. In 1933 Pounder was present at the sea trial of the reconditioned *Olympic* and he wrote *"It was fascinating to stand at the forward end of these magnificent engines, all their running mechanisms brightly polished and watch the enormous crossheads rising and falling harmonically, the valve gear swinging in rhythm, the great cranks turning silently and inexorably"*. This poetic description of these immense engines in motion brings out the intense love of large machinery, which dominated Pounder's life.

In 1932 he was appointed Chief Technical Engineer with responsibility for propulsion machinery, both steam turbines and diesel engines, becoming a Director in 1949. Up until the nineteen fifties there was a battle for supremacy between the steam turbine and the diesel engine. Pounder was active in the steam turbine field and was a member and then Chairman from 1951 of the Steam Turbine Committee of Parsons and the Marine Engineers Turbine Research and Development Association, PAMETRADA, until its demise in 1962, which virtually marked the end of steam turbines in marine propulsion. However there was no doubt that Pounder's real love was for the marine diesel engine.

Diesel patented his engine in 1892 and a successful prototype was tested in 1897. During the next decade MAN and Stork-Werkspoor demonstrated its application to ship propulsion. However real success came when the Danish company of Burmeister and Wain (B&W,) developed a direct-coupled reversible eight-cylinder four-stroke diesel engine of 2020 bhp, two of which were installed in the *Selandia* in 1911. This ship was lost off Japan in 1942. The B&W Oil Engine Co. Ltd. was formed in Glasgow in 1912 and in that year H&W took over part of the capital, acquiring the rest during the First World War. In 1917 they obtained the licence for the construction of B&W engines throughout the British Empire. By 1921 the technical staff and manufacture was transferred to the H&W Engine Works in Belfast under F.E. Rebbeck and Pounder became increasingly involved in the development of marine diesel engines.

Rudolf Diesel's prototype engine, 1893-96

Selandia, first ocean going diesel engine ship, 1912.

During the nineteen twenties and thirties there were many important and radical developments in the marine diesel engine. H&W, as by far the largest licensee of B&W, were closely associated with their implementation. Progressively Pounder became involved in these developments, taking over responsibility from F.E. Rebbeck who became Chairman of H&W in 1930. The original B&W diesel engines had air-blast injection of fuel into the cylinders which added considerably to the cost and complexity. Pounder, with Rebbeck, developed the much simpler system of solid injection. This was employed by Akroyd Stuart in his patented hot-bulb engine of 1890, which predated Diesel's patent. It became universally adopted for marine diesel engines including the engines of their licensor, B&W. Pounder was also involved in increasing the power output of marine diesels by developing pressure charging according to the Bachi system, using exhaust turbo-superchargers, and by under piston pressure-charging.

In the nineteen twenties the double-acting, four-stroke, crosshead-type engine was developed. 14,500 hp engines of this type were installed, for example in the 22,000 ton *Asturias* and *Alicantra* delivered in 1926 and 1927. Later the two-stroke diesel engine was developed and for larger powers the engine was a double-acting, crosshead-type. These were installed, for example, in the 25500 ton *Stirling Castle* and *Athlone Castle* delivered in 1934.

8 cylinder double-acting four-stroke engine with 680 mm bore diameter and 1600 mm stroke, developing 6000 ihp at 100 rev/min fitted to the 14,137 ton *Highland Monarch* launched 1928.

In its original form the double-acting, two-stroke engine with two exhaust pistons, top and bottom, of smaller diameter than the main piston posed some serious technical problems. It required a cylinder cover, top and bottom, which accommodated the two exhaust pistons, and the piston rod for the main piston passed through a seal in the lower exhaust piston. There were serious cracking problems in cylinder covers and it was difficult and costly to maintain the seal in the lower exhaust piston. In the late nineteen thirties, at the behest of Pounder, B&W redesigned the engine with exhaust pistons the same diameter as the main piston, which avoided the need for cylinder covers, but the problem of the piston rod seal remained.

6 cylinder double-acting two-stroke engine with 620 mm bore diameter and 1400 mm stroke, developing 6000 ihp at 98 rev/min fitted to the 11,122 ton *Australia Star* launched in 1935.

Up till the late nineteen thirties H&W had been largely dependent on B&W for the design of engines, though they contributed to the detailed design and they were responsible for some development work as well as the manufacture and installation. At that time Pounder got agreement from the Board of H&W to the need to become independent of B&W if necessary, as indeed arose during the war. As a consequence during the Second World War Pounder designed a single-acting, opposed-piston, two-stroke engine, which was first manufactured after the war. This was a radically simpler design than the double-acting two-stroke engine: it used simple materials, it was easy to produce, it was much easier to overhaul and maintain and it eliminated the cylinder covers.

The opposed-piston, two-stroke, crosshead-type engine with exhaust turbo supercharging was used extensively up till the retirement of Pounder in 1964 at the age of 73. The four-cylinder version of the engine with a bore diameter of 600m (24 in.) and 1800mm (71 in.) total stroke produced 4000 bhp, and the ten-cylinder version with a bore diameter of 750 mm (30 in.) and a total stroke of 2300 mm (91 in.) developed 25000 bhp. Besides the crosshead type engine there was also a more compact trunk-type engine. Engines were also supplied as stationary engines for electrical power generation (for example in the Channel Islands) and for pumping engines (for example the oil pipeline from Iraq to Tripoli), as well as for locomotives built by H&W for the Irish and South American railways.

With the retirement of Pounder in 1964 and the great decline in ship building in the UK, it became uneconomic to continue with the independent development of the H&W opposed-piston, two-stroke engine. H&W reverted to being completely dependent on B&W as a licensee. By that time B&W had developed a highly successful very simple poppet-valve, two-stroke, crosshead-type engine which was cheap to manufacture and reliable. During his long association with the Engine Works of H&W Pounder was responsible for the installation of 1,100-1,200 engines with an aggregate power of 6 million horsepower. He died on the 15 October 1982 at the age of 91.

Pounder liked classical music and he was an avid reader of Conan Doyle and Agatha Christie. He married Edith Winifred MacCauley in 1932, and according to his Personal Assistant, George Boyd, they married at eight o'clock in the morning and he was back at work at nine o'clock. He was an extremely literate man with a philosophical turn of mind, and he wrote many technical books and papers. He was a Fellow and one-time President of the Institute of Marine Engineers, and a Fellow of the Institution of Mechanical Engineers, the Royal Institution of Naval Architects and the Institution of Civil Engineers, and received numerous prize and awards.

Like many another 'Great Engineer' Pounder did not suffer fools gladly, and his relationships with some of his senior colleagues were notoriously poor, - in particular with Sir Frederick Rebbeck, Chairman of H & W. He had a contempt for academics, in particular Professors, and he had a fund of excellent stories of the stupidity of Professors. But on the other hand he sang the praises of his colleagues in the drawing office and of H&W, which he considered was the finest shipyard in the world, as indeed it was in his lifetime.

5 cylinder opposed-piston two-stroke turbo charged engine with 750 mm bore diameter and 2000 mm total stroke developing 7000 shp at 100 rev/min fitted to the 13,471 ton *Tresfenn* launched in 1960.

Sources:

1. 'Shipbuilders to the World – 125 years of Harland and Wolff, Belfast 1861 – 1986', by Michael Moss and John R. Hume, pub. The Blackstaff Press, 1986, ISBN 0 85640 343 1

2. 'Harland and Wolff – Burmeister and Wain Marine Diesel Engines and the Relevance of C.C. Pounder on the Development of the Two-stroke Marine Diesel Engine', by B. Crossland, pub. Trans.I. Mar. E., Vol. 98, paper 19, 1986.

3. Papers kindly lent by Mrs V. Pounder, daughter-in-law, and wife of the late Rafton Pounder MP.

Sir FREDERICK REBBECK

Frederick Ernest Rebbeck was born at Lockeridge near Swindon in Wiltshire on 19 August 1877, the son of Albert Ernest Rebbeck, farmer, by his wife Janette Smith. His father died in 1878 of tuberculosis. Frederick was educated privately before becoming an apprentice to the Metropolitan Amalgamated Railway Carriage and Wagon Company in the Midlands.

On the completion of his apprenticeship, in about 1897, he moved to Northern Ireland to broaden his experience by joining Victor Coates and Co in their Lagan Foundry. The origins of the company could be traced to 1799. It was one of the early manufacturers of steam engines, including an engine installed in the first steamship built in Belfast in 1820. By the end of the nineteenth century they were mainly concerned with building mill engines and steam engines for electrical power generation, including a 4000 hp engine for Newcastle upon Tyne. From 1901-03 he assisted in the teaching of science in the Royal Belfast Academical Institution. During this period he was awarded a Whitworth Exhibition. In 1903 he joined Harland and Wolff as manager of the engine work's drawing office, before leaving Belfast to join the British Westinghouse Company in Manchester, formerly the Metropolitan Amalgamated Railway Carriage and Wagon Company, which later became Metropolitan Vickers.

By the standards of his day he was well educated in engineering subjects, no doubt by evening classes after a working day, which probably started at 6 am. From 1893 at the age of fifteen through to 1902 when he was twenty-four, Rebbeck passed many national examinations set by the Department of Science and Art, which subsequently became the Board of Education. In many of his exams he passed as first class, which was awarded to very few.

He rejoined the engine works of H&W sometime before the First World War, but was seconded to Burmeister and Wain in Copenhagen. B&W had developed a four-stroke marine version of the diesel engine, which in 1912 was successfully demonstrated in the *M.V. Selandia*. This was the first ocean-going oil-engined merchant ship. It survived until 1942 when it sank after a collision with a submerged rock. In 1912 Lord Pirrie, Chairman of H&W, reached an agreement with Ivar Knudson of B&W to become their first UK licensee and to promote a new company –Burmeister and Wain (Diesel System) Oil Engine Company – which had been set up in the previous year. In July 1912, the new company purchased the Lancefield Engine Works at Finneston on the South side of the Clyde which H&W had purchased the previous year.

M.V. Selandia

In 1915 O.E. Jorgensen the manager appointed to run the Lancefield Works was replaced by Rebbeck. Three other Danish engineers, V. Mickelsen, the chief designer, his assistant J. Muller and A. Hammer the chief test engineer remained in position. All three were to remain with the company which was ultimately absorbed into the H&W engine works. Between 1915 and the end of the First World War ten marine diesel engine installations were built similar to the original installation in the *M.V. Selandia*.

In 1919 Lord Pirrie appointed Rebbeck as Managing Director of the engine works in Belfast while still managing the Lancefield Engine Works. Pirrie recognised a great future for the marine diesel engine and he planned that H&W would produce all the engines for the shipyards under his control. He also recognised the need for a reliable supplier of steel castings and forgings and, in 1919, Rebbeck was responsible for the planning and erection of what was to become the largest foundry in the UK, the Clyde Foundry, on a site in Helen Street, Glasgow.

The H&W Engine Assembly Works at Queen's Island.

To reduce costs Rebbeck was encouraged to introduce standard designs, and to transfer the design department in the Lancefield Works to Belfast in 1921.

C.C. Pounder, who had joined H&W in 1915 in the pipe layout office, joined the diesel engine team which he ultimately dominated. Though H&W was to remain a licensee of B&W, they were the largest manufacturers of their engines and the relationship was one of equals. Without research and development departments, engines installed in customer's vessels were virtually used as test beds for development. Rebbeck, with Pounder, was responsible for replacing the complex, and costly air blast fuel injection system used in the original diesel concept, by the much simpler and cheaper airless or solid injection system. This was adopted by B&W and other leading manufacturers of marine diesel engines.

When H&W financially supported B&W in the development of the more powerful double-acting, four-stroke, marine diesel engine, much of the supporting development work was carried out in Belfast. For example, Rebbeck and Mickelsen were responsible for designing the all important oil cooling of the piston rod and piston. The successful design and development of large double-acting four-stroke marine diesel engines provided the power for the 22,000 ton *Asturias*, 20,990 ton *Carnarvon Castle* and the 22,000 ton *Alcantara* delivered in 1926-27.

Shortly before Lord Pirrie's death in 1922 Rebbeck and Charles Payne, the shipyard manager, were placed in overall charge of the ship design department in addition to their normal duties. The death of Lord Pirrie, who had been chairman since 1896, left an enormous void. He was an autocrat who played his cards very close to his chest and kept financial matters within his London office. The Deputy Chairman, Lord Kylesant, a close confidante, took over as Chairman. Despite this, he was not aware of the perilous state of H&W's finances even though it was a member of the Royal Mail Group, a shipping conglomerate, that he chaired.

Lord Kylesant was faced with a formidable financial situation, made no easier by the senior manager's lack of financial responsibility arising from Pirrie's control of the money bags. This was not helped by the complete lack of cost accountancy and cost control. Large debts had accumulated as a result of the loan guarantees granted to H&W under the Trade Facilities Act. Added to this was the deteriorating world economic state in the nineteen-twenties. When the treasury and the Northern Ireland Department of Finance threatened foreclosing on the assets of H&W. Kylesant, like his predecessor, massaged the annual accounts which finally led to his resignation.

With some considerable hesitation the banks, the treasury and the Northern Ireland Department of Finance, finally agreed to the appointment of Rebbeck to the Chairmanship of H&W. He was seen as having little experience of financial control and he was essentially an engine man not a shipbuilder. It was a most unpropitious time to take over the reins of power with the onset of the great depression of the nineteen-thirties, and the resulting collapse of demand for ships. Ultimately after the completion of the *Georgic* in June 1932 the Belfast shipyard was closed and put on a care and maintenance basis and it was not reopened until Autumn 1933. During this time a few ships were built in the Glasgow yards of H&W and the engine works obtained some contracts for land based engines and diesel electric locomotives.

The Diesel Electric Locomotive on the Belfast Co. Down Railway

After 1933 there was a slow recovery in demand for ships, encouraged by vigorous advertising and promotion initiated by Rebbeck. There was also an order from the Admiralty for *HMS Penelope*, a 5,000 ton light cruiser, which was the beginning of the rearmament programme. But still there remained the problem of inadequate control of cost in the shipyard, though this was offset by profitable ship repair work at Liverpool, London and Southampton.

Rebbeck seized the opportunity provided by the 1933 Defence White paper to set up Short and Harland, a joint company, with Short Bros of Rochester. This was housed in a new factory alongside the Musgrave Channel and adjoining the proposed site of a Belfast airport. He also upgraded the toolroom in the Lancefield Works to support armament work, while the manufacture of gun mountings was expanded at the Scotoun Works on the Clyde. Later, again on Rebbeck's initiative, an agreement was reached for the War Office to erect an armaments factory in the Abercorn Works in Queen's Island.

Rebbeck's efforts in building up the company in the nineteen-thirties put it in a pre-eminent position to provide massive support to the UK's war effort. They were involved in a major programme of building for the navy including no less than ten aircraft carriers as well as specialised naval ships, while still building merchant ships and oil tankers. Aircraft manufacture was building up both within H&W and also in Short and Harland, while the Abercorn Works became involved in the manufacture of cruiser tanks 3.2 inch anti-aircraft guns etc. In recognition of his great contribution to the war effort Rebbeck was awarded a knighthood in 1941, and he received further recognition of a KBE in the Coronation Honours of 1951.

Mistakenly the Northern Ireland Government ignored the risk of air attacks, and they made no adequate provision for Civil Defence and Belfast was ill provided with anti-aircraft defences. They paid the price when on the 7-8 April 1941 a small squadron of German Bombers completely destroyed the H&W fuselage factory. This was the main supplier to Short and Harland, whose Rochester factory was nearly completely demolished later in the year. Much worse was to come on the 15-16 April when 180 bombers attacked Belfast, and probably by mistake heavily bombed the densely populated area north of the city centre. Probably 900 people or more were killed and many more injured with a horrendous loss of housing. It led to a mass exodus from the city and untold misery and demoralization. Then on 4-5 May there was a further massive raid, which caused immense damage to the industrial heartland of the city, though the death toll was much less at about 150.

Probably no city outside of London suffered so grievously as Belfast. Altogether H&W sustained damage assessed at over £3 million, an immense sum at that time, and the largest single claim on the Government during the war. The immediate effect was a substantial reduction in output to only a tenth of normal. Rebbeck was faced with a formidable task to reinstate the works, and he faced criticism for his determination on solidly built replacement buildings, which would be an asset in peace as well as war. It was during this period that the workforce dubbed him Lord Concrete. However, under his determined leadership H&W recovered to again make a major contribution to the war effort.

Rebbeck carried the immense burden of running a huge enterprise employing over 50,000 people in Belfast, Glasgow, London, Liverpool, Southampton and Londonderry. He was faced with satisfying a large variety of contracts, as well as introducing new

War time damage

technologies such as welding in fabrication. To cap it all he had to cope with reinstating the factory devastated by bombing and restoring the morale of the working force. It was not altogether surprising that there were serious financial problems as a result of delays in compensation for war damage and payment against the large number of contracts. Perhaps in the minds of the banks, Treasury and the Northern Ireland Department of Finance, Rebbeck did not give financial matters sufficient priority. The net result was that in 1943 there was a restructuring of the company and an independent Chairman was appointed, leaving Rebbeck to concentrate on the management of the H&W Belfast complex – a wise decision. Later in 1945 he was reappointed as Chairman.

Mainly as a result of Rebbeck's efforts and determination, H&W came out of the war in excellent financial fettle, and with modern and well equipped facilities. There was little serious competition and there was a great demand for new ships, and for the refurbishment of liners such as the *Capetown Castle*, which had been commandeered as troop ships. Though there were no new orders from the navy there was still much naval work in the completion of aircraft carriers under construction. The engine works under C.C. Pounder had been cut off from their licensor, B&W, during the war, and they had taken the opportunity to design an opposed-piston two-stroke marine engine, which was extensively adopted in the following two decades.

Under the autocratic management style of Rebbeck, following very much in the pattern of Lord Pirrie, H&W produced excellent financial results up to the end of the nineteen-fifties. They produced a large range of ships including liners, cargo liners, merchant ships, whale factory ships, whale catchers, oil tankers and bulk carriers. Besides marine diesel engines and steam turbines they also produced diesel engines and steam turbines for electricity generating stations as well as engine driven pumps for some of the long distance oil pipelines in the Middle East. They also built large reciprocating gas compressors for the petro-chemical industry under licence from the American company of Cooper-Bessemer Corporation, as well as radial flow compressors and large electric motors and generators.

The end of the nineteen-fifties was a period of great change in the ship building industry. As a result of the enormous increase in worldwide air travel there was a great decline in demand for passenger and cargo liners. Increasing world trade and the extensive use of containers for the transport of goods, led to the demand for ever larger bulk carriers, container ships and oil tankers. The new shipyards which had sprung up worldwide, and especially in the Far East, adopted radically new construction methods and fabrication processes. Welding completely supplanted rivetting for the fabrication of the hulls of ships. Although the 44,270 ton *Canberra* delivered in 1961 had a completely welded aluminium superstructure, there was still extensive rivetting in the fabrication of the hull. In the new technology very large pre-fabricated sections of ships were produced under cover in fabrication workshops, and then lifted into position in a building dock. On completion the vessel was floated out rather than launched from a slipway. H&W could not take advantage of this new technology as it had no building dock, nor the very large fabrication workshops and heavy lift capacity required, and this put them at a major disadvantage.

However British shipbuilders suffered another, and more serious problem, which did not afflict most of their competitors; in the complexity of their trade unions. In British shipyards, and H&W was no exception, the workforce was represented by a multiplicity of trade unions, which led to interminable inter-union strikes, some of which were extremely damaging. In most of the new shipyards there were single union workforces, which gave rise to far fewer disputes and more efficient use of labour.

Sir Frederick Rebbeck and his Secretary Miss Nancy Lowry

From 1958, after the very successful post-war financial performance, there was a decline due to the factors noted above and also losses incurred in the building of the *Canberra*. In 1961 Rebbeck fell ill and he finally retired in March 1962 in his eighty-fifth year. He died on the 27 June 1964 at the age of eighty-six.

Sir Frederick Rebbeck had proved to be a man of iron during his very long service in management of H&W. His autocratic style of management, learned at the feet of Lord Pirrie, had proved to be a boon during the great depression of the early nineteen-thirties and the trauma of the war years. But perhaps in the end when he was over seventy he did not sufficiently recognise the immense changes taking place in the shipping world, and the revolution taking place in shipbuilding technology. Though his relationship with the workforce in H&W was better than appertained in other shipyards, nevertheless he proved unable to overcome the stranglehold of the trade unions. There was the added problem of Government confusion in their support of strategic industries such as shipbuilding. Though enormous changes were brought about after his retirement, it proved too late to bring about the radical changes in time to give the yard a long term chance of survival. At the time of writing, March 2003, the last ship to be built by H&W has sailed from Belfast.

Besides his management role in H&W, Sir Frederick was a director of several other companies, including Short Bros and Harland. However he was also interested more broadly in matters relating to shipbuilding and engineering. Besides being a member of numerous engineering institutions he served as President of the Institute of Marine Engineers in 1931, President of the Shipbuilding Employers' Federation in 1936/37, and Vice-President of the Royal Institution of Naval Architects 1940-61. For 33 years he served on the Technical Committee of Lloyds Register of Shipping and was Chairman or Joint Chairman for 20 years. He was a member of the Belfast Harbour Commissioners. In 1944 he received an Honorary DSc from the Queen's University of Belfast.

In 1907 he married Amelia Letitia, daughter of Robert Glover of Chester, Cheshire, who predeceased him in 1955. They had two sons and four daughters. One of his sons, Dr Denis Rebbeck, became Deputy Managing Director H&W in 1955 and Managing Director on his father's retirement in 1962, becoming Chairman in 1965.

Sources:

1. 'Shipbuilders to the World; 125 Years of Harland & Wolff, Belfast, 1861-1896', by M. Moss and J.R. Hume, pub. The Blackstaff Press 1986. ISBN 0 85640 343 1

2. 'A History of Ulster', by Jonathan Bardon, pub. The Blackstaff Press, 1986, ISBN 085640 466 7

3. 'The Verdict of Peace', Chapter II 'This Long and Bitter Dispute: the Shipyards', by Correlli Barnett, pub. McMillan, 2001, ISBN 0 333 67982 2.

4. 'British Shipbuilding: The Harland and Wolff Yards, Belfast', by E. Cuthbert and D. Rebbeck, pub. Welding, Sept. 1947

JOHN STEVENSON

John Stevenson was born in Rostrevor Co. Down in 1850, eldest son of John and Harriet Stevenson (nee Walsh) and died 30 May 1931. While he was still a child, the Stevenson family moved to Belfast where John Stevenson Sr., was a book-keeper. He was educated at the Royal Belfast Academical Institution.

In 1871, Stevenson set up business in partnership with a Mr Campbell. This firm, Stevenson & Campbell, printers, designers and lithographers, had premises in Upper Arthur Street. In the mid 1870s this partnership was dissolved and Stevenson was joined by Mr Alexander McCaw and Mr James P. Orr to form McCaw, Stevenson and Orr in April 1876. Business expanded rapidly and soon the premises became too small. In August 1878 the partnership purchased new premises at 16 - 18 Linenhall Street, from a firm of Linen Merchants, Gordon Brothers Ltd. From then on the works were called "The Linenhall Works".

Alexander McCaw had spent some time in America prior to returning to Belfast. It is possible that he had acquired some infection there because his health started to deteriorate and he died in November 1880. Both his sons were taken into the firm but as the elder, George was then only twenty years old, he could not take up a directorship. Alexander's widow looked after the family's interest and took a keen interest in the business. In 1882 George was sent back to the United States to develop the market for the firm's products. Having established a branch there he returned to set up a London office and became a full partner in 1885.

As well as the normal printed products, two major products of the firm were Stevenson's patented ideas. These were "Glacier" window transparencies and "Seccotine" glue. The former was patented as number 1874 of 1880 "Transparent paper or cloth labels for advertising" and included the names of all three partners with Stevenson listed first.

"Glacier" windows were stained glass look-alike sheets in which the design was lithographed onto a film and attached to plain glass. It could be used as an advertising medium in a shop window or as a cheap way of creating a stained glass window in a private house. Stevenson also came up with a method of stiffening the conventional display cards which were normally printed as single sheets of card. After a very short time these would normally curl up and become unsightly. The novel framing process which he

Typical examples of the "Glacier" windows

adopted, using another layer which would curl the opposite way, prevented this. Over the years, Stevenson made a significant number of other patent applications in relation to printing processes. However, the majority were not taken beyond the first stage either because he had come up with a further improvement or because there was not enough commercial benefit from proceeding.

The period at the end of the eighties and early nineties saw a further considerable expansion of the firm and was increasingly profitable for it. Turnover increased from just over £5,000 in 1878 through £11,615 in the financial year ending in April 1882 to £30,315 in 1885. A factor of aproximately 100 should be applied to make comparisons with present values. As the firm expanded, they acquired more property by purchasing adjacent buildings in Linenhall Street and leasing 19 - 23 Franklin Place, which abutted the existing factory. As a result of the success which the firm enjoyed, the partners decided, in 1889, to become a limited liability company with a nominal capital of £50,000.

By the end of the decade the premises were again too small for the volume of business and the directors entered negotiations to purchase a large site on the Stranmillis Road, where they would construct a purpose built factory and housing for managers and foremen. A row of terrace of houses was built and named Chillworth Buildings and Ridgeway Street laid out but the factory itself did not go ahead. Instead, the Loopland Mill, at that time the premises of the County Down Flax Spinning Company, on the Castlereagh Road was purchased and this provided excellent facilities for expansion. The firm moved there in 1894 and made full use of the four storied building and the surrounding single story top lit former weaving sheds which provided a total of about 50,000 square feet. In 1896 the company employed some 800 workers in the new premises and turnover rose to £75,600.

At the time, as well as McCaw, Stevenson & Orr, Belfast had another internationally famous printing firm, Marcus Ward & Co. Ltd., which had been founded in the late 1840s. It occupied the Royal Ulster works on the corner of Bankmore Street and Dublin Road. The company was famous for the quality of its chromo-lithographic work and was one of the earliest mass producers of Christmas and other greetings cards. Its other staple product was the range of Vere Foster educational books designed for the teaching of writing and drawing. The two sons of Marcus Ward were unable to work together and Vere Foster placed his work with John Ward when he broke away from the family firm in 1876. This led to a long series of law suits which ended up in favour of Vere Foster and resulted in Marcus Ward Ltd., going into liquidation in 1899. Stevenson acted quickly and bought the assets of Marcus

Ward & Co. Ltd. in January 1900. The products included leaf calendars, block calendars and the "Automatic" copying letter book for making copies of hand-written letters. In 1904, the firm started making the boxes for a new Liverpool concern "Meccano", manufacturers of the construction system which enthralled generations of children. McCaw, Stevenson & Orr, also took on the work for "Hornby" Trains when Meccano developed that line. The business connection lasted well into the 1970s.

During the First World War the requirement for general printing fell off and the demand for "Glacier" products, which dominated the company's foreign business, dried up almost completely. However, the demand for "Seccotine" increased quite considerably. It was of invaluable use to aircraft manufacturers and a testimonial was received from Short Brothers, who at that time were based in Eastchurch.

"We have now for some considerable time past used SECCOTINE in connection with the construction of our aeroplanes, and have pleasure in testifying that this has given us every satisfaction. We do not know of any other adhesive that would have served our purpose."
SHORT BROS.

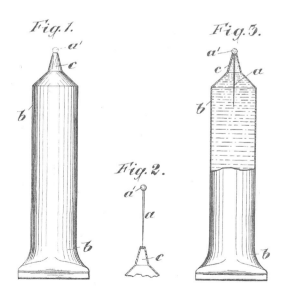

The Patent Drawings for Stevenson's Glue Tube

"Seccotine" was a multi-purpose, fish based, glue. Both it and the re-sealable tube in which it was sold were Stevenson's ideas. He had been experimenting with glues as far back as the late 1870's when he had rented separate premises in Sandy Row for the manufacture of glues. Over the period of twenty odd years he had developed his eventual formula and registered the trade name under which it was first marketed in 1896. However the simplicity of his idea, which was granted a patent (3128 of 1908), for using a pin first to create a hole in the neck of the flexible zinc alloy tube in which the glue was sold and then seal the hole, can be considered one of those strokes of genius which mark the true innovator.

As a result of his experiments with the different materials from which glues can be made, Stevenson developed a bizarre range of other products. These were marketed by the company using their own trade names. They included "MAR-AKKA" toothpaste, "CAPE" metal polish and "RISTONA" gold, silver and lustre paints.

As well as his investigations into the problems of chemistry and chemicals, Stevenson took an interest in physics. During 1889, he tried to get Harland & Wolff sufficiently interested in the idea of underwater transmission of sound to have them assist experiments. The idea was the basis of a patent application 6011 of 1892 "Audible Marine Signalling Apparatus". It was not taken any further at the time but the principle he described in a letter to E. H. Harland was later developed as sonar which enables the tracking of submarines.

In the 1920's considerable changes were introduced in the litho printing trade. The massive stones used to that time were being replaced by flexible zinc plates and camera and film technology were advancing enabling four colour separations to be employed. The combination of different processes including offset printing gave excellent results from watercolour originals but could not be patented as such. McCaw, Stevenson & Orr gave it the trade name "Macrotype" a contraction of McCaw Rotary Type.

Stevenson was not simply a printer who advanced the business through inventions and modernisation of traditional skills. He was a man of extremely wide-ranging interests. Unlike the conventional Ulster businessman of the period as well as being an inventor, he was an antiquarian, archaeologist, author, linguist, poet and traveller. During the 1890s he was the editor and publisher of a weekly periodical "The Pen." From his own contributions to this paper he compiled his first book *Pat "MaCarty,*

Farmer, of Antrim - his rhymes with a setting." This was published by Edward Arnold of London in 1903. It was followed by A Boy in the Country also published by Edward Arnold in 1912. Stevenson then translated *De Latocnaye's Promenade d'un Francais dans l'Irlande, 1796.* This was not taken up by any of the London publishers who were approached during 1914 and 1915 and he eventually published it himself in 1917 as *A Frenchman's Walk through Ireland.* His major work Two Centuries of Life in Down, published in 1920 by McCaw, Stevenson & Orr and by Hodges, Figgis & Co., in Dublin, was a substantial tome of historical research and must have been a labour of love to such a busy man.

Stevenson was rewarded for his literary achievements by the award of an Honorary Master of Arts degree from Queen's University of Belfast in 1926.

John Stevenson died on 29 May 1931. His widow, Catherine Ross Stevenson continued to take a close interest in the firm until her death after the Second World War. Their son Leslie (1885 - 1957) continued as a director and chairman until his death.

Sources

1   'From Linenhall to Loopbridge', by Bryan McCabe, pub. MSO, Belfast 1990

2   Belfast Newsletter

3   Belfast Telegraph

JAMES THOMPSON

James Thomson was born on the 16 February 1822 at College Square East in Belfast, in a house opposite to the Royal Belfast Academical Institution. He was the third child of a family of seven, of which the fourth son, William, born on the 26 June 1824 was to become the world famous scientist Lord Kelvin.

College Square East; the house to left of lamp-post was the home of James Thomson Snr.

He was descended from a long line of farmers who migrated from Scotland during the Plantation of Ulster in the early seventeenth century. His father, also James, was born on the family farm at Ballymaglave near Ballynahinch where he was brought up as a farm labourer. Though he only had a rudimentary home education, he developed a remarkable inventive and mathematical ability. This led his father to allow him to attend Dr Edgar's school at Ballynahinch where he soon made himself useful as an assistant teacher. From 1809 until 1814 he studied at the University of Glasgow taking his degree of MA in 1812. On completion of his studies in 1814 he was appointed as a teacher in the School Department of the newly created Belfast Academical Institution which in 1821 became the Royal Belfast Academical Institution. A year later, in 1815, he was appointed Professor of Mathematics in the College Department which provided the first higher education courses in Ireland outside of Trinity College, Dublin.

In 1817 James married Margaret Gardner who died in 1830 a year after the birth of their youngest son. Then in 1832 James was appointed Professor of Mathematics in the University of Glasgow. He was much involved in the education of his family and in particular James and William, who remained great friends and collaborators throughout their lives. Both of them at a very young age benefited from attending lectures in the University and in 1834, James then twelve and William aged ten, became students in the University. During their University education William was normally first in the class followed closely by James in second place. In 1840 James, at the age of eighteen, was awarded an MA in Mathematics and Natural Philosophy.

On leaving the University of Glasgow in 1840, James determined to become a Civil Engineer and, to gain practical experience he joined the office of Mr McNeill in Dublin. McNeill, later Sir John McNeill, became the first Professor of Practical Engineering in Trinity College, Dublin in 1842, only the second chair of engineering in the UK. However after a few weeks James took ill and was forced to return to Glasgow where he attended the lectures of the first Professor of Engineering, D. B. Gordon, appointed in 1841.

In 1842 his health having improved he continued with getting experience firstly with a family friend, Mr J.A. McClean at Walsall, followed by a spell in the famous Horsley Iron Works at Tipton in Staffordshire. His father then paid £100 for an apprenticeship with Mr Fairbairn, later Sir William Fairbairn, at his Engine Works at Millwall and then at Fairbairn's Manchester Works where he met and worked for Mr William Fairbairn, one of the great Victorian engineers. But at the end of 1844 he became seriously ill and was forced to return home where the family doctor advised him that his condition was life threatening. As a consequence he was constrained to confine himself to work which did not physically tire him.

The next four or five years were one of the most formative periods of his life, during which he was to lay the foundations of his many considerable contributions to invention and science. Though he never published extensively, everything he wrote was of seminal importance to engineers and scientists and he took extraordinary care in the clarity of exposition. Though he got on extremely well with his brother, Lord Kelvin, and they frequently collaborated, their approaches were very different. Without disrespect, Lord Kelvin published at the drop of a hat. In his lifetime he published 686 papers and books together with 70 patents; the more remarkable as for much of his life there was no accepted or commercially available typewriter. James's publications in his lifetime are contained in a modest single volume of his collected works.

In 1849 there was an outbreak of cholera in Glasgow, not an unusual occurrence in the large cities of the UK at that time, caused by inadequate sanitation. James Thomson Senior caught cholera and died on the 12 January 1849 at the age of 62. James who had by that time recovered his health appears to have been completely devastated by his father's death. He went to London for a period where he appears to have seriously questioned his religious faith but ultimately came out of it with his faith greatly strengthened. Then in 1851 his elder sister, Anna Bottomley, who lived in Belfast persuaded him to settle in Belfast and practise as a Civil Engineer. She introduced him to the firm of J&J Herdman of Sion Mills where he was able to install a prototype vortex turbine to power their spinning mill and prove the system.

In Belfast James set up an office and soon attracted worthwhile clientele in a town which, at that time, was rapidly expanding into what was to become one of the great manufacturing cities in the UK. Besides continuing his engineering and scientific interest he became deeply involved in the intellectual life of Belfast. He was also concerned with the great social problems of his day and age and, in 1852, he read a paper to the newly formed Belfast Social Inquiry Society entitled 'On Public Parks in Connexion with Large Towns'. This led directly to the acquisition of the large Ormeau Park for the citizens of Belfast. Later, in 1869, he read a paper

'Nationalisation of Public Works' to the Belfast Engineering and Architectural Association. He advocated that the railways, which extended throughout the entire country, should belong to the state while local water works and gas works, with their associated distribution networks, should belong to the local community. At that time these were revolutionary ideas though, ultimately, they were accepted and implemented. Only in more recent times have these enterprises been re-privatised.

In 1853 he was appointed Resident Engineer to the Belfast Water Commissioners a post he occupied till 1857 when he was appointed to the Chair of Civil Engineering in the Queen's College of Belfast. The Chair had been vacated by Professor John Godwin, who had been appointed in 1849, when the department was created. As recorded by Professor John Perry, one of James's students, all the students had a loving memory of him as distinct from a mere liking or mere respect. He went to inordinate lengths to help students to complete their courses. Then, in 1872, Professor Macquorn Rankine, a professor of great eminence, died, and the Chair of Civil Engineering in the University of Glasgow became vacant. With encouragement from his brother, now Sir William, James applied and was appointed to the Chair in 1873. In Glasgow he continued with his many interests till 1888 when his sight began to fail and he retired from his chair in 1889.

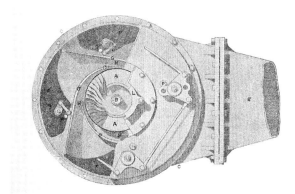

Plan View of Vortex Turbine.

James's contributions to engineering and science were of great significance and it is only possible in the compass of this article to mention briefly some of them. Perhaps his greatest contribution was in the field of fluid mechanics. In 1846 he invented the inward flow whirlpool or vortex turbine which, with encouragement from

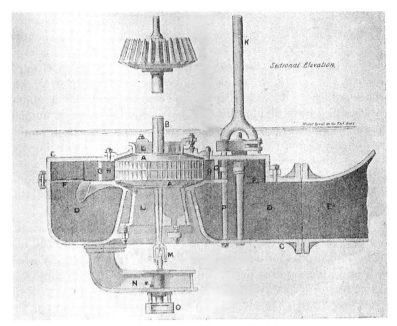

Transfer Section of Vortex Turbine.

his brother, he patented in 1850. The design included the very novel idea of pivoted inlet guide vanes which were controlled by an external linkage. This allowed the flow of water into the turbine to be controlled, and hence the output power to be varied. The principal licensee was Messrs Williamson and Bros of Kendal which became the internationally well recognised company of Gilkes and Gordon. The excellence of the design was recognised by the award of a medal of merit in the 1862 South Kensington International Exhibition, the follow up of the Great Exhibition of 1851.

Water turbines designed by James were extensively exploited including in Ireland. The basic design has been extensively applied up to the modern day for large, power generating, water turbines. However, they are mostly referred to as Francis Turbines, after J.B. Francis who contemporaneously developed a similar turbine with fixed guide vanes. This does not recognise the primacy of James's contribution.

The need to measure accurately the power output of his vortex turbine led him to investigate the design and performance of friction brake dynamometers which have been used ever since for testing power producing machines. The principle of the vortex turbine led him to invent the centrifugal pump and fan. In 1855 he designed large centrifugal pumps for use in drainage and irrigation in Jamaica and later in 1856 the design of a very large pump for drainage of a large tract of land below sea level in Demerara in British Guiana. Land drainage led him in 1852 to design the very simple jet pump, which has no moving parts and only needs a high head supply of water.

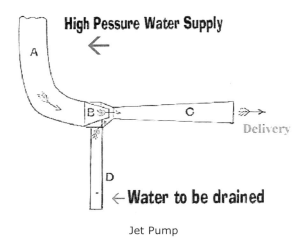

Jet Pump

During his period as Resident Engineer of the Belfast Water Commissioners he became interested in the measurement of water flow in rivers and streams. In particular he investigated the use and performance of v-shaped notches as being superior to rectangular notches. He also investigated the windings of rivers in alluvial plains and the flow of water round bends in pipes.

Perhaps his most fundamental contribution was in 1847 when he applied the principle expounded by Sadi Carnot in 1824, known as the Carnot's cycle, which is one of the corner stones of thermodynamics, to the lowering of the temperature of the melting point of ice with pressure. Later in 1873 when considering the relation between the gaseous, the liquid, and the solid state of water, he noted the existence of the 'triple point' at which all three states co-exist. James's work demonstrated to Clausius, Rankine and Lord Kelvin an approach which led them to their development of our understanding of thermodynamics.

Another subject of great importance to which James devoted much thought was the safety of engineering structures and the principles on which their sufficiency in strength could be estimated and proved. When he became interested in the subject in 1862 he felt that tests usually applied were insufficient to permit an engineer feeling justified in risking the lives of men and the property of their employers. He considered that severe overload tests should be applied and refuted the belief that it would weaken the structure if it had been correctly designed. Safety was the subject of his inaugural address given in Glasgow in 1873; needless to say presented in Latin. The principles he enunciated were the basis of engineering practice up till fairly recent times, though with increasing complexity it is now recognised they are not sufficient.

Early in 1868 he became interested in geology and in particular the Parallel Roads, Shelves or Terraces of Lockaber. There had been much debate on how these roads had been created and in a paper in 1848 James provided a glacial explanation which satisfied all the geological data and was widely accepted. In 1877 he also wrote a definitive paper 'On the Jointed Prismatic Structure of Basaltic Rocks' which provided an explanation of the formation of the Giant's Causeway on the Northern Coast of Country Antrim.

Giant's Causeway

James Thomson Sr. in 1846 wrote a new edition of his book 'Geography' and he asked his son to look into the question of trade winds and to write a short account of this atmospheric phenomena. His dissatisfaction with the, then, stated theories aroused his life long interest in the subject. After retirement and despite the deterioration in his eyesight, but with the help of his wife and youngest daughter, he devoted his remaining years to the subject. This culminated in his Royal Society Bakerian Lecture - he had been elected a Fellow in 1877 - on the 'Great Currents of Atmospheric Circulation', which was presented to the Society on the 10 March 1892.

In the following May James, his wife and youngest daughter all caught a severe cold which was quickly followed by inflammation of the lungs - what would probably now be called pneumonia. James died on the 8 May 1892 followed by his youngest daughter three days later and his wife a week later.

Finally it is worth quoting Dr J.T. Bottomley's final paragraph of his epitaph on his brother-in-law, James Thomson:

> Purity and honour in word and deed and thought, gentleness of disposition, readiness to spend his labour, his time, his mental energies for others, and for the good of the world in general, all were conspicuous in his life both in public and in private.

Sources:

1. 'The Life of Lord Kelvin' by Silvanus P. Thompson, pub. McMillan & Co., London, 1910, 2 vols, pp 1297.

2. 'Collected Papers in Physics and Engineering' by James Thomson, DSc, LLD, FRS, edited by Sir Joseph Larmor, DSc, LLD, SecRS, MP, pub Cambridge University Press, 1912, pp 484.

3. 'On the Vortex Water-Wheel' by James Thomson, pub. Report of the British Association for the Advancement of Science 1852, pp 317, 322.

WILLIAM THOMPSON
LORD KELVIN OF LARGS

William Thomson, 1852

Most people regard Lord Kelvin as one of the greatest scientists of the nineteenth century, and they might question his inclusion amongst the 'The Great Engineers of Ulster'. However, though Kelvin's contribution to our understanding of the fundamentals of science are legendary, nevertheless he also applied his science to many practical problems as represented by the seventy patents granted to him in his lifetime. He was also elected the first President of the Institution of Electrical Engineers when it was formed in 1888, out of the earlier Society of Telegraph Engineers. In 1892 he was elected President of the Institute of Marine Engineers and, from 1874, he was a member of the Institution of Civil Engineers. So it must be recognised that he was as much an engineer as a scientist and more than worthy of inclusion.

William Thomson was the fourth child of a family of seven and was born in Belfast on the 24 June 1824. His elder brother, James, born two years earlier, was nearly as famous as William but as an engineer with a fundamental science bent. Their family background and life up to leaving Glasgow University has been told in the essay on James Thomson so there is no need to repeat it here.

From 1834, at the age of ten, when he matriculated, William studied in the University of Glasgow excelling in every subject he took. He was no slouch and, like many students of his time, he studied a wide range of subjects across the arts and sciences; not the narrow specialism of present day degree courses. His studies, for example, included such subjects as humanities, natural history, Greek, Latin, logic, mathematics and natural philosophy. In 1841, at the age of seventeen, he left university without taking a degree. Though he had passed his examinations he abstained from a formal conferment. He had already published his first paper "On Fourier's Expansion of Functions in Trignometrical Series" in the Cambridge Mathematical Journal.

In 1841, Thomson, still only seventeen, formally entered St Peter's College, Cambridge, now more normally known as Peterhouse, where he studied under several eminent academics for the next four years. Though his main interest was in mathematics and natural philosophy he took advantage of the opportunity to attend lectures in other subjects and even toyed with the idea of switching his interest to the law, which held better prospects. As was to be expected he took part in the social and cultural life of his college including rowing, playing music, and attending cultural evenings with the dons. In the examinations in January 1845, he was disappointed by only achieving the position of Second Wrangler, whereas his friends had confidently expected him to be the Senior Wrangler. However, shortly after he was awarded the prestigious First Smith's Prize.

He then spent a few months in Paris where he devoted much of his time studying experimental methods in research into natural philosophy of which he had no experience at that time. On his return to Cambridge he became an assistant examiner and took on students to coach in mathematics. In June 1845 he was elected a Foundation Fellow at £200 per annum.

He and his father envisaged the likely death of Professor Meikleham, who held the Chair of Natural Philosophy in Glasgow, and the opportunity this might provide for William to be elected to the vacant chair, though at the time he was only twenty one. Professor Meikleham died in May 1846, and, on the 11 September, William was elected to the Chair. He gave his inaugural lecture in Latin on the theme "Age of the earth and its limitations as determined from the Distribution and Movement of Heat within it", on the 13 October. He was to remain at the University of Glasgow until his death on the 17 December 1907, being made a Knight Bachelor in 1866 and Baron Kelvin of Largs in 1882.

On entering his duties as Professor of Natural Philosophy on the 1 November 1846 he read an introductory lecture on the scope and methods of physical science which he was to deliver for over fifty years though he did not rigidly adhere to it. It was expected that the new professor would develop the teaching of experimental physics which had virtually been neglected prior to his appointment. In 1846 he inherited equipment, much of it over one hundred year's old and little less than fifty years old. In his first year a grant of £100, not an insignificant sum at the time was provided to purchase instruments and subsequently more funds were made available. This was the beginning of William Thomson's development of experimental research needed to provide the physical data for his theoretical researches.

At that period of history there were no such appointments as research students and post-doctoral research fellows and professors carried out their own research work. Thomson, when he had obtained some modern research equipment and the use of a disused wine cellar, started experimental research. His only support was from amongst his students who were excited by his infectious enthusiasm. This laboratory must be accounted the first working laboratory for physical science, though there were earlier private laboratories created by individuals for their own use.

In the eighteen fifties, his laboratory established, Thomson turned his attention to the practical application of science which was to make him internationally famous as well as wealthy. Electric telegraphy had been developed in the late eighteen thirties and by 1850 overland telegraphy was well established and a profitable business. It depended on the discovery of gutta-percha which proved to be an excellent insulating material for the copper wires. By 1849 short lengths of submarine cables had been laid and in 1851 a successful cable was laid from Dover to Calais. The problem which attracted William's attention was the distortion of the electrical signals which was a serious potential limitation as the length of cable increased. Its effect was to reduce the rate of transmission which would have made a trans-Atlantic cable uneconomic.

In 1854 William applied Fourier's mathematics to the problem of signal distortion in electric telegraphy and he showed that the distortion of the signal was in direct proportion to the 'capacity' and 'resistance' of the cable. This led to a Royal Society paper "On Practical Methods for the Rapid Signalling by Electric Telegraphy", in 1858, in which he explains a proposed system for a submarine cable between Ireland and Newfoundland. Rapid signalling was essential to ensure the economic success of such a cable.

In October 1856 the Atlantic Telegraph Company was formed with Professor William Thomson nominated by the Scottish shareholders as one of eighteen directors. Mr O.E.W. Whitehouse was initially responsible for the electrical instrumentation to test the insulation and continuity of the cable and to transmit and receive messages. However Thomson played a pivotal role in the ultimate success of the enterprise. He established the importance of the purity of the copper used for the cable in reducing resistance, and he developed the electrical instrumentation for testing the cable and for transmitting and receiving messages which was ultimately widely adopted. When cable failure occurred while laying the first cable, he developed improved paying out gear which he patented. He was also present and actively involved in all the cable laying leading up to and including the final successful operation.

The British Government provided the HMS *Agamemnon* and the United States Government lent the US frigate *Niagara*, which were both fitted out with hurriedly devised cable paying out gear. The two ships assembled at Queenstown in the South of Ireland, now known as Cobh, and then sailed to Valencia Bay in the far South West of Ireland and the nearest point to Newfoundland. It was planned that the *Niagara* would lay the first half of the cable and, in mid Atlantic, the cable end would be spliced to the cable being carried by the *Agamemnon* which would complete the laying to Newfoundland. The shore end of the cable was landed at Valencia on the 5 August 1857 and after covering 330 nautical miles the cable parted in water 2000 fathoms deep. This brought the operation to an end for 1857.

An additional length of cable was manufactured and a new paying out mechanism was designed by Thomson for the next attempt. In May 1858 successful trials were carried out in the Bay of Biscay, with Thomson on board the *Agamemnon* with his recently manufactured mirror galvanometer. For the second attempt it was decided that the two ships would proceed to mid-Atlantic, splice the two cable ends, and then proceed to pay out in both directions. On the 24 June the splice was made, and paying out commenced, but after several failures the two vessels lost contact and returned to Queenstown. On the 17 July they again departed and on the 29 July they spliced the ends in mid-Atlantic. They both managed to reach land on the 5 August, completing the first trans-Atlantic cable, and telegraphic contact was made immediately. Unfortunately the cable only worked for a few weeks, probably due to the application of too high a voltage by Whitehouse. The only successful transmission of message was achieved using Thomson's equipment. Whitehouse disgraced himself and was dismissed.

It was not until 1864 that further plans were agreed to attempt the laying of a trans-Atlantic cable. In the meantime there had been considerable scientific advances, such as the adoption of units of electrical measurement, refinement of the means of electrical measurement and the use of a signalling condenser to sharpen the electrical pulses which augmented the speed of signalling. In 1864 capital was raised to construct and lay a new cable. As a result of Thomson's work it was agreed to greatly increase the cable size to 300 lb of high conductivity copper and 400 lb of gutta-percha per nautical mile. With iron wire armouring this gave a total weight of 1.8 ton/mile. It was a particularly opportune time as Brunel's Leviathan, the 22500 ton *Great Eastern*, was available to lay the cable. With both screw and paddles it was very manoeuvrable and also capable of holding the entire cable.

The *Great Eastern* leaving Shearness with the telegraph cable on board.

On the 23 July 1865 the *Great Eastern* departed from Valencia. At 1250 miles the cable parted in 2100 fathom depth of water. Though they recovered the cable three times they lost it on each occasion because of the failure of a swivel bolt on the gear used to raise the grapnels.

The cable laying gear on the *Great Eastern*

On the 13 July 1866 the *Great Eastern* picked up a new shore end at Valencia and spliced it to a new cable. It proceeded to Heart's Content Bay in Newfoundland arriving on the 22 July without a significant hitch. Communication had been maintained with Valencia via the cable throughout the voyage. Within twenty four hours of completion the line was inundated with messages from Europe. Subsequently the *Great Eastern* returned and recovered the end of the cable laid the previous year on the 2 September. Having spliced it to the remaining cable on board they successfully returned to Newfoundland, so the company then had two perfect cables.

Receiving Messages from the *Great Eastern* Instrument room at Valencia Island.

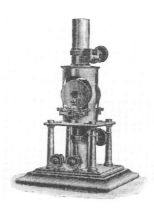

Thomson's Marine Mirror Galvanometer, 1858.

The success of this great enterprise and the important role paid by Thomson led to Queen Victoria awarding him a knighthood as a token of her appreciation of his service in connection with the Atlantic Telegraph and in recognition of his high position in science and classical attainment. He was knighted at Windsor Castle on the 20 November 1866. He received world wide recognition of his achievements and many honours.

For many years Sir William remained actively associated with major projects for other submarine cables. This experience together with a life long interest in ocean sailing in his yacht, the 126 ton *Lalla Rookh*, created an interest in navigation. This included an investigation into compasses leading to the Kelvin's compass widely used in the maritime world and by the British Navy. When involved in cable laying he had been concerned with the ineffective method for taking soundings so he developed a sounding machine which allowed soundings to be taken much more quickly and easily. He also gave consideration to lighthouse lights, tide-predicting machines and tables for facilitating the use of Sumner's method of finding the position of a ship at sea. His knowledge on maritime matters led to his serving on Admiralty Committees on the Design of Ships of War in 1871 and Types of Warships in 1904 - when he was eighty years old.

Despite his involvement in important engineering works during much of his life, Sir William continued to make major fundamental contributions to many branches of science, as reported in many of the 686 books and papers he published in his lifetime. With Professor Peter Guthrie Tait he wrote the classical "Treatise on Natural Philosophy" published in 1867. This marked an epoch in

William Thomson, Lord Kelvin, with his compass, 1902

the teaching of the foundation of physical science which at the time many considered scarcely less important than Newton's "Principia Mathematica".

On New Year's Day 1892 it was announced that a peerage of the realm had been conferred on Sir William Thomson by Queen Victoria. The Prime Minister, Lord Salisbury, in making the offer expressed the anticipation that his influence on the House of Lords as an eminent man of science would greatly strengthen their deliberations. On the 23 February Sir William Thomson was gazetted as Baron Kelvin of Largs in the county of Ayre.

William Thomson married Margaret Crum in 1852 but for much of their married life she suffered ill-health dying in 1870. In 1873 while involved in the laying of a submarine cable from Lisbon to Brazil, the specially built cable laying vessel, the *Hopper*, was anchored off Madeira for a couple of weeks. There he met Frances Anna Blandy whom he married in Madeira in 1874.

During his long life Lord Kelvin published no less than 686 books and papers and was granted 70 patents which made him a wealthy man. Up to within a few weeks of his death at the age of eighty three on the 12 December 1907 he was still carrying out experimental work. As a measure of the esteem in which he was held by the public, he was buried in Westminster Abbey on the 23 December next to the grave of Sir Isaac Newton.

Sources:

1. 'The Life of Lord Kelvin' by Silvanus P. Thomson, pub. McMillan & Co., London, 1910, 2 vols, pp 1297.

2. Illustrated London News

Sponsored by Professor John McCanny, C.B.E., F.R.S., F.R.Eng., School of Electrical and Electronic Engineering, The Queen's University of Belfast.

WILLIAM ACHESON TRAILL

William Acheson Traill was born in 1844, the second son of William Traill of Ballylough House, Bushmills Co. Antrim. He was educated privately and then at Trinity College, Dublin, where he graduated as a Civil Engineer. He succeeded in developing and running the first hydro-electric railway in the world at the Giants Causeway Co. Antrim. The solutions to the problems he faced in creating this system formed the basis of virtually all the electric tramways introduced in towns and cities throughout Britain and Europe.

The unique, basalt rock formation known as the Giant's Causeway is one of the natural wonders of the world. Because of its isolated location, it was described by the famous Dr Samuel Johnston as being "worth seeing but not worth going to see". The nearby Bushmills distillery had been licenced since 1608. The Victorian interest in sea bathing and the growing affluence of the industrial classes led to the development of the village of Portrush into a thriving seaside resort. It was first connected to Belfast by the Ballymena and Portrush Railway company in 1855. This company and its successor, the Belfast and Northern Counties Railway, considered extending their line to the Giant's Causeway to take full advantage of the tourist potential. From 1846, Irish railways could only be built to the standard gauge of 5' 3 " and the expense of such track outweighed the possible increase in profits. However, the passing of the Tramways Act in 1870 facilitated the construction of roadside tramways of narrower gauge. These were mainly urban lines powered by horses but a scheme was proposed to form the Ulster Steam Tramway Company to run a steam powered tram from Portrush to the Giant's Causeway. Its promoters were unable to raise adequate finance and it never materialised.

Following his graduation, Traill was employed by the Geological Survey and, among other projects, had surveyed the Iron Ore and Limestone deposits along the north coast of Antrim. With his brother, Dr Anthony Traill, he proposed a railway to link the Causeway with Dervock, on the Ballycastle Railway which was then under construction and a roadside tramway to link up with the BNCR main line at Portrush. They were sufficiently forward looking to ensure that their Act included the possibility of using animal, mechanical or electrical power on any of the proposed lines. They received the Royal Assent to the Act in 1880 and then faced the problem of raising finance. Their contacts through Trinity College only succeeded in attracting some £20,000 to meet the estimated cost of constructing the seven miles of roadside tramway.

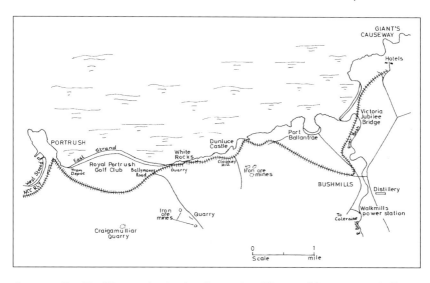

Among Dr Traill's contacts in the scientific world was Dr William Siemens, one of the German brothers who were pioneers in the generation and use of electricity. Siemens Brothers Ltd., the English part of this early multi-national group, took £3,500 of shares in the Giant's Causeway tramway and Dr Siemens was given a seat on the board. At the 1879 Berlin Exhibition, the Siemens brothers had demonstrated a working electric railway using their patented dynamo and motor. Other small demonstrations had been staged around the world but the Traills' project, at six miles, lifted it into the realm of a full-scale, public service tramway.

The ceremony of cutting the first sod for the tramway was held on 21 September 1881 and performed by Mrs W. A. Traill. A steam powered generator was built at the Portrush terminal to augment the main source of power, hydro-electricity, which was to be generated at Walkmills on the River Bush near Bushmills. The system was designed to work at 225/250 volts D. C. As track laying proceeded, Dr Edward Hopkinson, Siemens representative on the site, carried out preliminary tests of the electrical systems. The earlier Siemens railways had employed the two rails as conductors for the current. On the North Antrim coast the inadequacies of the system became obvious very rapidly. In dry weather the cars could operate satisfactorily up to two miles from the generating station. However in the normally salt laden and moist atmosphere encountered on site, current leakage proved too great. The use of a third rail alongside the hedge and raised about seventeen inches above ground level helped slightly but leakage still prevented the carriages from operating more than a couple of miles from the generators. Traill then patented a new version in which the conductor rail could be attached to either top or bottom of a supporting beam and used improved insulation materials for the supporting posts. Where the power conductor required an opening across a road junction or field gateway, the rails were bonded by an underground copper wire wrapped in insulating

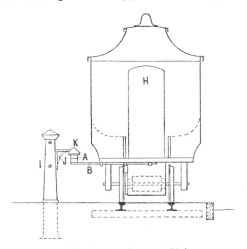

Traill's Patent Current Pickup

Sources:

1. 'The Life of Lord Kelvin' by Silvanus P. Thomson, pub. McMillan & Co., London, 1910, 2 vols, pp 1297.

2. Illustrated London News

Sponsored by Professor John McCanny, C.B.E., F.R.S., F.R.Eng., School of Electrical and Electronic Engineering, The Queen's University of Belfast.

WILLIAM ACHESON TRAILL

William Acheson Traill was born in 1844, the second son of William Traill of Ballylough House, Bushmills Co. Antrim. He was educated privately and then at Trinity College, Dublin, where he graduated as a Civil Engineer. He succeeded in developing and running the first hydro-electric railway in the world at the Giants Causeway Co. Antrim. The solutions to the problems he faced in creating this system formed the basis of virtually all the electric tramways introduced in towns and cities throughout Britain and Europe.

The unique, basalt rock formation known as the Giant's Causeway is one of the natural wonders of the world. Because of its isolated location, it was described by the famous Dr Samuel Johnston as being "worth seeing but not worth going to see". The nearby Bushmills distillery had been licenced since 1608. The Victorian interest in sea bathing and the growing affluence of the industrial classes led to the development of the village of Portrush into a thriving seaside resort. It was first connected to Belfast by the Ballymena and Portrush Railway company in 1855. This company and its successor, the Belfast and Northern Counties Railway, considered extending their line to the Giant's Causeway to take full advantage of the tourist potential. From 1846, Irish railways could only be built to the standard gauge of 5' 3 " and the expense of such track outweighed the possible increase in profits. However, the passing of the Tramways Act in 1870 facilitated the construction of roadside tramways of narrower gauge. These were mainly urban lines powered by horses but a scheme was proposed to form the Ulster Steam Tramway Company to run a steam powered tram from Portrush to the Giant's Causeway. Its promoters were unable to raise adequate finance and it never materialised.

Following his graduation, Traill was employed by the Geological Survey and, among other projects, had surveyed the Iron Ore and Limestone deposits along the north coast of Antrim. With his brother, Dr Anthony Traill, he proposed a railway to link the Causeway with Dervock, on the Ballycastle Railway which was then under construction and a roadside tramway to link up with the BNCR main line at Portrush. They were sufficiently forward looking to ensure that their Act included the possibility of using animal, mechanical or electrical power on any of the proposed lines. They received the Royal Assent to the Act in 1880 and then faced the problem of raising finance. Their contacts through Trinity College only succeeded in attracting some £20,000 to meet the estimated cost of constructing the seven miles of roadside tramway.

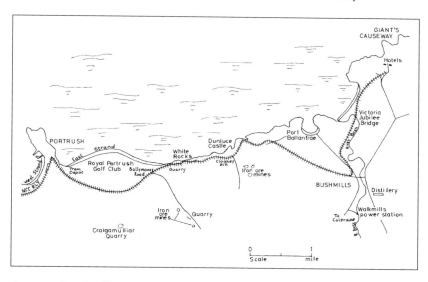

Among Dr Traill's contacts in the scientific world was Dr William Siemens, one of the German brothers who were pioneers in the generation and use of electricity. Siemens Brothers Ltd., the English part of this early multi-national group, took £3,500 of shares in the Giant's Causeway tramway and Dr Siemens was given a seat on the board. At the 1879 Berlin Exhibition, the Siemens brothers had demonstrated a working electric railway using their patented dynamo and motor. Other small demonstrations had been staged around the world but the Traills' project, at six miles, lifted it into the realm of a full-scale, public service tramway.

The ceremony of cutting the first sod for the tramway was held on 21 September 1881 and performed by Mrs W. A. Traill. A steam powered generator was built at the Portrush terminal to augment the main source of power, hydro-electricity, which was to be generated at Walkmills on the River Bush near Bushmills. The system was designed to work at 225/250 volts D. C. As track laying proceeded, Dr Edward Hopkinson, Siemens representative on the site, carried out preliminary tests of the electrical systems. The earlier Siemens railways had employed the two rails as conductors for the current. On the North Antrim coast the inadequacies of the system became obvious very rapidly. In dry weather the cars could operate satisfactorily up to two miles from the generating station. However in the normally salt laden and moist atmosphere encountered on site, current leakage proved too great. The use of a third rail alongside the hedge and raised about seventeen inches above ground level helped slightly but leakage still prevented the carriages from operating more than a couple of miles from the generators. Traill then patented a new version in which the conductor rail could be attached to either top or bottom of a supporting beam and used improved insulation materials for the supporting posts. Where the power conductor required an opening across a road junction or field gateway, the rails were bonded by an underground copper wire wrapped in insulating

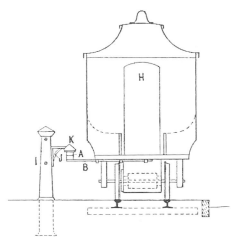

Traill's Patent Current Pickup

material. Traill also patented an underground conduit for the third rail power supply which was adopted by several cities around the world.

Initially it had been intended to use a battery in each passenger vehicle to improve their performance at a distance from the generator and give some freedom away from the third rail. This proved unacceptable due to weight considerations. Two of the open first class cars were provided with electric motors the two other first class coaches and three open third class cars were to be unpowered trailers as were the goods and mineral wagons. While the electrical connections and operational details were being finalised two steam locomotives were purchased to allow the Board of Trade inspection to be carried out on 12 January 1883 and the start of public operations on 29 January 1883. Due to legal problems, the construction of the hydro-electric power station could not be started until March 1883. However, the steam generator provided enough electrical energy at the Portrush end of the line for the BoT inspectors to permit its use over the whole length.

William Traill was sufficiently proud of his tramway system, the first electric tramway in the United Kingdom and the first in the world to be powered by hydro-electricity, to request that it should be opened by Queen Victoria. Negotiations on this idea delayed the official opening and it was not until 28 September 1883 that the Viceroy, Earl Spencer, formally declared the tramway open.

The tramway proved an immediate success and the board decided to increase its length to eight and a half miles by extending the line to the Giant's Causeway as their initial prospectus had intended. This project included the construction of a new station at Bushmills and a bridge across the River Bush called the Victoria Jubilee Bridge. which was opened on 1 July 1887. The terminus at the Causeway had a remarkable building made of corrugated iron with a gracefully curved, high pitched roof which served as booking office and waiting room. The trackwork allowed for a long run around loop, each side of which could accommodate rakes of eight vehicles - a very optimistic move by Traill as the electric motor cars had difficulty pulling three trailers. About this time it was decided that the goods and mineral traffic was not up to the expected level and five goods wagons were converted to passenger carrying configuration.

This increase in traffic also demonstrated that the original dynamos and motors were already outdated. A new generator for the

Walkmills hydro-electric station was obtained from Ewell, Parker Ltd. This enabled the voltage to be raised on occasions to counteract the leakage from underground cables linking the generator to the rails. However things got to such a state that Traill decided to use only overhead power lines and re-equipped the system.

In August 1895, a cyclist attempting to turn his bicycle on the road, fell across the tram track and ended up with his shoulder on the third rail. He died half an hour later. It is therefore unlikely that his death was as a result of electrocution. At the opening of the railway, Lord Kelvin, the eminent scientist who had invested £1000 in the venture, had gone around shaking hands with unsuspecting guests while holding a live wire in his other hand. He believed that such electric shocks were good for many illnesses. He had also demonstrated the safety of the system by dropping his trousers and sitting on the live conductor rail. However, a Board of Trade inspector carrying out an investigation after the fatal accident found many regulations were being ignored and that the voltage could be increased to over 300 volts. His report instructed the company to limit the voltage to 250 and raise the live rail to at least eight feet above rail level or find some alternative means of current supply. The voltage restriction meant that only one electric car could use the track at a time. A new steam locomotive was bought and the service maintained, at a cost.

The Tramway in Operation.

Quotations were obtained for the conversion from third rail to an overhead trolley system but these proved too expensive for the company. Traill then resigned his seat on the board and acted as contractor to carry out the conversion by direct labour. Two new toast rack motor-cars with overhead trolleys were obtained and the change over was effected in July 1899 without interruption to the traffic on the tramway. Traill was allotted £5,000 in debentures and £3,000 in preference shares for the work. He was then co-opted back on to the board. When the necessary Board of Trade approval had been received, a new generator was ordered and the supply raised to 550 volts. With new cars being ordered the output was inadequate and a further generator was delivered in 1907. These new motor-cars permitted the use of three car rakes and therefore catered better for the tourist season. Steady improvements were made until 1911 including a new gas powered generator for the Portrush depot to be used in case of a drought.

In May 1916, the Causeway Tramway became the only British railway to be damaged by naval action during the war. The "Wheatear", an coaster en route from Coleraine to the UK was confronted by a German U-boat which surfaced and attempted to capture her off Portballintrae. As the coaster was armed shots were exchanged one of which put a twelve foot crater beside the track and another cut a straining wire for the overhead poles and two off duty employees had their clothes cut by flying splinters.

After the war the tramway system continued to provide a service for tourists but Traill started to worry about the supply of water in the river Bush. He therefore added a new diesel powered generator at the Portrush Depot in 1925. Although supplied by the same manufacturer as the dynamo at Waklmills, the two could not work in parallel. This problem was solved by installing a switch at Dunluce to divide the system and having each generator supply one portion when the demand was high. At quiet times either generator could supply the whole system. The two steam-powered tram engines were last used in that year and later sold. The open toast-rack cars received roof canopies in 1936 following an incident where the trolley boom fell on an open car, fortunately without fatal consequences.

William Traill married three times. In 1879, his first wife was Harriet Jane daughter of Barton Wrigley of Hoylake, Cheshire. She died in 1885. His second, whom he married in 1891, was Elizabeth Jane daughter of Edward Graham of Ivy Lodge, Newry. His third whom he married in 1925 was Norah, daughter of Robert Proctor Woodhouse of Preston. He had met her in unusual circumstances when he had saved her from drowning when the boat in which she and her parents had been sightseeing had capsized in the summer of 1895. Traill died on 6 July 1933 and was succeeded as Chairman of the Tramway company by his third wife. The company struggled through the Second World War, assisted by fuel rationing for private cars and the presence in Ulster of American servicemen who found the experience of riding on the unique railway a way of using their spending power. However, the realities of post-war Britain forced final closure on 30 September 1949.

Sources

1. 'The Giants Causeway Tramway', by John McGuigan, The Oakwood Press, 1964

2. 'Giants Causeway, Portrush & Bush Valley Railway & Tramway Co. Limited', by John McGuigan, UFTM, 1983

GUSTAV WILHELM WOLFF

Gustav Wilhelm Wolff was born in Hamburg on the 19 November 1834, the son of F. Moritz Wolff a respected Jewish merchant. Perhaps more important for his future his mother, Fanny Maria (née Schwabe), was the sister of Gustav Christian Schwabe whose wife was the sister of Edward Harland's aunt. Gustav Schwabe was to have great influence on the lives of both Edward Harland and Gustav Wolff, and on the development of Harland and Wolff.

At the age of fourteen Wolff, having been privately educated in Hamburg, was sent to Liverpool where his uncle, Gustav Schwabe, had his business and was also a director of J. Bibby, Sons and Company. He attended the Middle School of the Liverpool Collegiate Institution for one year. The College had opened a few years earlier, and the Middle School provided a broad education for boys intending to enter commerce and trade, which included mathematics, drawing, chemistry and natural philosophy. In 1851 he took up an apprenticeship with the world renowned engineering enterprise of Joseph Whitworth and Company based in Manchester. On the completion of his apprenticeship he represented his company on their stand at the Paris Exhibition of 1855, the successor to the Great Exhibition held in London in 1851.

After a period as draughtsman with B. Goodfellow Ltd of Hyde on the outskirts of Manchester, he became a personal assistant to Edward Harland in the Robert Hickson's shipyard in Belfast in 1857. This was the beginning of his life long association with Belfast and the shipyard which was to become Harland and Wolff.

Some time after joining Robert Hickson's shipyard Wolff was sent to gain engineering experience at sea, a practice which was to be encouraged subsequently for many engineers trained in the shipyard. Experience at sea was seen as a vitally important element of training. When Edward Harland took over the yard in 1858 he recalled Wolff to take charge of the drawing office. Gustav Schwabe had advised Harland to take over Hickson's yard and he was instrumental in the shipping company of J. Bibby, Sons and Company placing an order for three 1500 gross ton steamships. This necessitated the extension of the yard and additional new machinery.

Bibbys were well satisfied with their new steamships and in 1860 they ordered six more vessels of 1,800-2,000 gross tons. This involved further expansion of the yard and the need for new capital. In 1861 Edward Harland and Gustav Wolff entered into partnership to form Harland and Wolff (H&W). Not only did Wolff bring managerial and engineering support to the enterprise, but through his wealthy family he had access to financial expertise and resources essential to the development of the business.

In 1868 Gustav Schwabe helped Thomas Henry Ismay in acquiring the White Star Line and subsequently to set up the Oceanic Steam Navigation Company. This led to the new company ordering six transatlantic steamships of 3,800 gross tons. Though in Wolff's obituary in the Belfast Evening Telegraph of the 17 April 1913 it is stated that he took the first order for the White Star Line, this ignores the role of Gustav Schwabe and of Edward Harland. What has to be recognised is the close collaboration of the family in many successful ventures; it was a case of all for one or one for all.

The (2nd) *Oceanic*

The Oceanic, the first of the steamships ordered by the Oceanic Steam Navigation Company, made its maiden transatlantic voyage in March 1871. It was, by the standards of the day, a sumptuously appointed ship and it can be considered as the first passenger liner providing real comfort. Wolff was still associated with H&W as a principal in 1899, when the second *Oceanic* of 17,274 gross ton was delivered, which was the largest vessel to have been built worldwide up to that time. The association with the White Star Line culminated with the construction of the three sister vessels - the *Olympic*, *Titanic* and *Britannic* delivered in 1911, 1912 and 1915 respectively, with gross tonnage of 45,324-48,158.

With the increasing size of the business the partnership was reconstructed in 1873 to embrace William J. Pirrie, Walter H. Wilson and Alexander B. Wilson who subsequently resigned in 1875 to pursue other business interests. By that time both Edward Harland and Gustav Wolff had begun to develop other interests. Gustav Wolff had become a partner in the newly created Belfast Ropework Company and by 1874 he had become Chairman, a post which he retained for many years. His association with this company gave H&W the competitive edge in the supply of rope and sailcloth required for their steamships, which still had sails, and also for the many sailing vessels they continued to construct up till 1890. The ropeworks also provided employment for many wives and daughters of H&W employees.

Belfast Ropeworks

The Belfast Ropework Company originally used machinery imported from America but later they developed and manufactured their own machines using castings produced by the local foundries. Gradually machinery for producing rope and cordage from jute was made by the local textile machinery manufacturers and particularly James Mackie and Sons. This led to the pre-eminence of Mackies in

The main rope-walk of the Belfast Ropeworks.

developing and marketing a complete range of jute machinery. During the Chairmanship of Gustav Wolff the Belfast Ropework Company became the largest ropeworks in the world. With the decline of demand and inadequate management, the company went into terminal decline closing in the second half of the twentieth century, to make place for the Connswater Industrial and Shopping Complex.

During the heyday of the Belfast Ropework Company, Gustav Wolff in 1887-88 carried out an extensive tour of India with Thomas Ismay and his wife. Ostensibly the tour was to negotiate for the supply of jute and sisal for the works back in Belfast but perhaps more to satisfy his inveterate love of travel and to indulge his favourite recreations of shooting and fishing. During this tour the story is told that he complained of the excessive price of the ticket for an overnight sleeper to Alahabad only to find that, as a result, the railway staff uncoupled the coach and left him at Cawnpore.

In 1884 both Gustav Wolff and Edward Harland ceased being partners who shared in the management of the company. But they remained principals and continued to make substantial loans to the company when called on. Though Wolff had withdrawn from day-to-day management, he continued to assist the partners in the direction of the business and on financial policy. As late as 1896, when William Pirrie was pre-occupied with his civic duties as Lord Mayor of Belfast, Wolff devoted considerable time to assisting the management on the direction of the business. In 1906, when

William Pirrie was created Baron Pirrie of Belfast in the Birthday Honours, Gustav Wolff announced his retirement and he sold his shares to Pirrie.

Edward Harland had become a Conservative MP for North Belfast in 1889, and in 1891 Gustav Wolff was elected Conservative MP for East Belfast, the heartland of H&W employees. He was re-elected in 1895, 1900, 1906 and January 1910 but when there was another election in November 1910 he announced his retirement from politics at the age of seventy-six. The involvement of senior industrialists in local and national politics might seem surprising in this day and age but at the end of the nineteenth and beginning of the twentieth centuries Home Rule for Ireland was perceived as a threat to Ulster. The fear of industrialists and their predominantly Protestant work force was that Home Rule for Ireland would lead to full independence which it was considered would be a serious threat to industry in the North of Ireland. The partners of H&W were so concerned that serious consideration was given to relocating their business to the mainland.

When Gustav Wolff retired from politics his constituents demonstrated their appreciation of his long service as their MP, by entertaining him and making a presentation. Besides his long service as an MP he had also been very generous in supporting good causes, such as the Ulster Hospital for Women and Children in Templemore Avenue and the Royal Victoria Hospital. He had done so much for Belfast and was so greatly liked and admired that the Corporation unanimously voted him the Freedom of the City in February 1911.

Gustav Wilhelm Wolff died on the 17th April 1913 at the age of seventy-eight in his London residence. He had remained a bachelor all his life.

Sources:

1. Death notice for Gustav Wilhelm Wolff, Belfast Evening Telegraph, 17 April 1913.

2. 'Shipbuilders to the World - 125 years of Harland and Wolff 1861-1986', by Michael Moss and John R. Hume, pub. The Blackstaff Press, 1986. ISBN 0 85640 343 1.

3. L.N. Lightbody, Bursar and Clerk to the Governors of Liverpool College provided information on the Liverpool Collegiate Institution attended by G.W. Wolff.

FRANCIS WORKMAN

Frank Workman was born in 1856. He was the fourteenth and youngest child of Robert Workman, a linen merchant, of Ceara, Windsor Avenue, Belfast and his wife Jane Service. The young Frank did not want to go into the linen business like his elder brothers, so he served his time as a premium apprentice at the shipbuilders Harland & Wolff. It is possible that he was influenced by the marriage of his elder sister, Mary, to George Smith of the City Line.

The firm of Workman, Clark & Co. was founded by Frank Workman in 1879. On first appearances, he seems to have set up his operations in the former premises of A. McLaine & Sons, whose business and goodwill had been taken over by Harland & Wolff in 1878 on the death of the last member of the McLaine family. He was joined by George S. Clark in 1880, also a former premium apprentice with Harland & Wolff, and, in 1891, by Charles E. Allen, whose family owned the Allen Line. Both Clark and Allen came from Scotland.

The original yard covered about 4 acres on the north bank of the river Lagan, to the north of the Milewater basin. In the first year four vessels totalling about 4,000 tons were constructed by a workforce of 400 men. The business expanded rapidly and around 1890 an additional yard was obtained on the south bank of the river adjacent to Harland & Wolff on Queen's Island. Until 1891 the marine engines for ships constructed by the firm were imported from Scotland. In that year their Victoria Engine Works and boiler shops were added to the firm's new south bank premises and Charles E. Allen was brought in to manage them. In 1892, the 5,050-ton *Southern Cross* was their first vessel to be fitted with Belfast built engines. It had been hoped that the owners, Wincott, Cooper and Co., would place further orders but these did not materialise. A related company, Houlder Bros., did place several orders later in the decade.

With the construction of the 3,239 ton *Star of Victoria* in 1887, Workman Clark introduced specially designed cooled cargo carriers to the meat trade. These were operated by a London based subsidiary of the Belfast firm of J. P. Corry & Co. It was followed by the 3,511 ton *Star of England* in 1889 and the 4,712 ton *Star of New Zealand* in 1892. They also built the 3,737 ton refrigerated carrier *Celtic King* for Wm. Ross &Co., in 1890. Earlier demonstrations of the technique for carrying frozen meat had used converted vessels and partially opened the British market to the Australian and New Zealand sheep farmers. This had an immense impact on, and benefit for, the consumer. No longer had the meat to be canned or salted for preservation on the long journey from farm to market. Co-incidentally it was in Belfast, New Zealand, that the first commercially successful freezing plant was constructed for this trade. They also built three sister ships of 3,300 tons, the *San Jose*, *Limon* and *Esparta* for The United Fruit Company in 1904 as the first wholly cooled fruit carriers for the South American trade.

In 1893 Workman Clark purchased the older established firm of McIlwaine and McCall, which, as McIlwaine & Lewis, had taken over the Ulster Iron Works, founded in 1836, and expanded from general engineering and repair work into shipbuilding. This site was across Queen's Road from their Victoria Engine Works and gave them five building berths on the south side of the river. In 1896 a disastrous fire which started in the nearby woodworking area of their neighbour, Harland and Wolff, destroyed the engine works and much of the shipyard. The firm rapidly re-built their facilities and took the opportunity to modernise their plant.

The list of clients for Workman Clark included most of the major shipping companies of the world at the period. As well as the City Line, which was operated by a brother in law of Frank Workman, and the Allan Line, the family business of Charles Allan, the firm built for Alfred Holt Ltd (known as the Blue Funnel Line), the China Mutual Steam Navigation Company, the Cunard Steamship Co., the Hamburg & America Line, the Norddeutscher Lloyd, the Ocean

Steamship Co. and the West India & Pacific. For many years at the turn of the century they featured in the top ten British shipyards in terms of tonnage launched. In 1904 the total tonnage launched at 44,272, was 12,430 greater than that of Harland & Wolff.

Workman Clark produced vessels which were complimentary to the output of their larger neighbour Harland & Wolff by concentrating on cargo liners with turbine engines. They held a licence for the production of Parsons steam turbine engines and, in 1903, constructed the 10,630 ton passenger vessel *Victorian* for the Allan line, the first transatlantic liner with this type of propulsive power. The Cunard company was so impressed with the performance of the vessel and the sister ship *Virginian*, built by Stephens on the Clyde, that they decided to build two more of this type for their own North Atlantic service. Unfortunately for Workman Clark, these vessels, the *Lusitania* and the *Mauretania*, were built by John Brown and Swan Hunter respectively. In due course Harland & Wolff built *Olympic*, *Titanic* and *Britannic* for the White Star Line with a combination of reciprocating and turbine engines to compete with these luxurious transatlantic liners.

The Allan Line *Victorian* on the slip in 1903. Comparisons should be made between the scaffolding employed here and H&W gantries of the same date.

After the declaration of the First World War, Workman Clark continued to build some merchant vessels including *Ebro*, and *Essequibo*, each of 8463 tons, for the Pacific Steam Navigation Co., the former was the ship on which Lord Pirrie died. They were also allocated war work. Among other ships six fast patrol; boats of 470 tons and 4,000 horsepower, sixteen boom defence vessels and four stern-wheel hospital vessels. Later in the war, when the government decided to build standard ships to replace tonnage lost to the German U-boats, Workman Clark were included in the firms given such contracts. They achieved records for riveting and were noted for the speed at which they built these vessels without reducing their quality. The Belfast Newsletter reported in 1919 that on Tuesday 10 September 1918 the ship launched from Queen's Island in the morning had its engines and boilers placed on board and installation work completed by nightfall. Final mooring trials were completed by Friday 13 September and the ship was ready for sea by late the following evening.

Turbines for the *Victorian* in the Engine Shop.

Frank Workman married Sarah McCausland (b 1865) and had a son and a daughter. Edward McCausland Workman died of his wounds in France in January 1916. In his memory, Frank Workman made a large donation to the Presbyterian War Memorial Hostel and took an active part in its development and management. Frank Workman died on 14 November 1927 and his widow on 10 April 1932.

He devoted most of his time and energy to the shipbuilding business and only entered public life in 1908 when he was elected to the Belfast City Council as a Unionist councillor representing Cromac ward. That he quickly earned the respect of other councillors is shown by his being elected High Sheriff in 1913. He also served as High Sheriff for County Down in 1917.

Following the First World War, the Workman Clark shipyard suffered the same fate as many competitors throughout Britain. There was a short surge of orders to replace lost tonnage and, as world conditions stagnated, the lack of new orders led to financial trouble and forced a reorganisation into Workman Clark (1928) Ltd. A government-backed organisation, The National Shipbuilders Security Limited (NSS), was formed to buy redundant shipyards and dismantle or mothball them. To avoid falling into their hands the management of Workman Clark (1928) approached Harland & Wolff in 1930 with a view to a merger. H&W had their own problems and postponed making any hasty decision. NSS were applying pressure to H&W to reduce the extent of their under-used assets on the Clyde. In April 1935, NSS took control of Workman Clark (1928) and a deal was struck whereby H&W took over the Workman Clark sites on Queen's Island and relinquished some of their facilities on the Clyde to the NSS for dismantling.

Sources

1   'Shipbuilders to the World - 125 Years of Harland & Wolff, Belfast, 1861 - 1986' by Michael Moss and John R. Hume, pub The Blackstaff Press Belfast, 1986. ISBN 0 85640 343 1

2   The Belfast Evening Telegraph

INDEX TO ILLUSTRATIONS

The illustrations in this volume come from a number of sources. Those listed as from "Authors' Archives" indicate that they were given to one or other author by the original copyright holder, at some time in the past, for their personal use in any form.

Author :- 6.3; 8.4, 8.5;
Authors' Archives :- 1.3, 1.4; 5.3, 5.4; 7.1, 7.2, 7.3, 7.4, 7.5, 7.6; 8.1, 8.2, 8.3;9.1, 9.2, 9.3, 9.4, 9.5, 9.6, 9.7; 11.2; 12.2, 12.4, 12.5, 12.6; 15.6; 16.2, 16.3, 16.4, 16.5; 18.2, 18.3,18.4, 18.5, 18.6; 19.2, 19.3,
W. & G. Baird Ltd. :- 6.1; 20.1, 20.2, 20.3
Belfast Natural History and Philosophical Society Centenary Volume :- 3.1; 15.1; 21.2
Belfast Industrial Heritage Trust Ltd. :- 1.4; 10.3; 14.5; 23.4
Belfast Telegraph :- 25.1
Robin Cameron :- 4.1, 4,3
Collected Papers in Physics and Engineering :- 21.1, 21.3, 21.4, 21.5, 21.6
Roger Fenton :- 2.1
Norah Ferguson :- 2.2; 6.2; 23.2
Raymond Hall :- 19.1, 19.4, 19.7
Howden Sirocco Ltd. :- 7.6
Illustrated London News :- 2.3, 2.4, 2.5; 22.2, 22.3, 22.4;
Life of Lord Kelvin :- 22.1, 22.5, 22.6
Gordon Mackie :- 12.1, 12.2, 12.3
Martin Baker Ltd. :- 13.1, 13.2, 13.3, 13.4
John McGuigan :- 23.4
Patent Office :- 20.4; 23.3
Valerie Pounder :- 18.1
Proceedings of the ICE Vol. 7 :- 15.2, 15.3, 15.4, 15.5
Queen's University of Belfast :- 11.1, 11.3
RIAC Archives :- 8.6
Arthur Stanley :- 14.1, 14.2
UFTM :- 3.2, 3.3, 3.4; 5.1, 5.2; 16.1, 16.6, 16.7; 23.1
UFTM (Harland & Wolff Collection) :- 1.1, 1.2; 4.2, 4.4, 4.5, 4.6; 10.1, 10.2, 10.4; 17.1, 17.2, 17.3, 17.4, 17.5; 19.5, 19.6; 24.1, 24.2
UFTM (Workman Clark Archive) :- 25.2, 25.3

The Lives of Great Engineers of Ulster